Raising Hell,
Living Well

Raising Hell, Living Well

Freedom from Influence in a World
Where Everyone Wants Something from You *

Jessica Elefante

*Including Me

BALLANTINE BOOKS | NEW YORK

Published in the United States by Ballantine Books, an imprint of Random House, a division of Penguin Random House LLC, New York.

BALLANTINE is a registered trademark and the colophon is a trademark of Penguin Random House LLC.

Library of Congress Cataloging-in-Publication Data
Names: Elefante, Jessica, author.
Title: Raising hell, living well: freedom from influence in a world where everyone wants something from you / Jessica Elefante.
Description: New York: Ballantine Group, [2023]
Identifiers: LCCN 2023000641 (print) | LCCN 2023000642 (ebook) |
ISBN 9780593500552 (hardcover) | ISBN 9780593500569 (ebook)
Subjects: LCSH: Self-actualization (Psychology) | Resilience (Personality trait) |
Conduct of life.
Classification: LCC BF637.S4 E53 2023 (print) | LCC BF637.S4 (ebook) |
DDC 158.1—dc23/eng/20230415
LC record available at https://lccn.loc.gov/2023000641
LC ebook record available at https://lccn.loc.gov/2023000642

Printed in Canada on acid-free paper

randomhousebooks.com

2 4 6 8 9 7 5 3 1

First Edition

Book design by Diane Hobbing

—For Hays and Beau—

It's said that the juice is worth the squeeze. As luck would have it, I've found that the best parts of life live up to that saying. This book, the two of them, and the long way round to both are proof of just that.

Contents

Part III:
The Bad Influence

Part IV:
Not So Easily Influenced

Part V:
Influential

Part VI:
Above the Influence

I will not mislead you anymore. I will not mislead you anymore.

I will not mislead you anymore. I will not mislead you anymore.

I will not mislead you anymore. I will not mislead you anymore.

I will not mislead you anymore. I will not mislead you anymore.

I will not mislead you anymore. I will not mislead you anymore.

I will not mislead you anymore. I will not mislead you anymore.

I will not mislead you anymore. I will not mislead you anymore.

I will not mislead you anymore. I will not mislead you anymore.

I will not mislead you anymore. I will not mislead you anymore.

I will not mislead you anymore. I will not mislead you anymore.

Apology Letter

Dear Reader,

I have, for most of my life, been a bullshit artist. A charlatan, a huckster, a trickster, a scoundrel, a cheat. Or better put: a brand strategist, a salesperson, a marketing executive, a bartender, a communications director, a thought leader. For a long time, at the expense of not only our society's well-being, but also of my own values, I influenced people to buy into whatever I was hawking.

Please forgive me. I'm not actually a terrible person. At least I don't think I am. In fact, I'm overly honest and a terrible fucking liar. But like you and all the other people living under the influences of our culture, I was conned into seeing reality in a certain way.

That's where this book comes in. Consider it my (mostly repentant) penance. I don't want to mislead you—anymore.

A Scatterbrain's Note on Being Precise

(or a Note on Names, Timelines, & Privacy)

Growing up I was given the nickname *Mother Goose* because of the tales I would tell. Like any good storyteller, I'm sometimes prone toward exaggeration. We fablers know that a highly colored *folklore*—or tale—is one remembered.

On that note, some people don't want their tales to be told. So in places where it mattered, I have changed names, locations, or situations to protect the person who would like to remain anonymous, and at times have created composite characters when I thought it was best. In some instances, I've consciously melded moments or decided to be vague with dates out of respect for someone's privacy. Where I wrote about my professional work, I didn't want to speak on behalf of others, so I have said "I" even when it included very talented and creative co-workers, founders, teams, agencies, freelancers, influencers, artists, colleagues, writers, producers, collaborators, or employees. And to obscure the categories, clients, and brands, I have mishmashed, stirred up, and compounded them.

I take many things very seriously. Life and I? Not too seriously at all. As an example, everyone in my family was really upset to discover that for the past three years we've been celebrating our dog's birthday on the wrong day. I didn't see the big deal—he still got a celebration.

For me, years, dates, time, numbers, ages, locations, people, and memories all whirl around, jumbled up, and misplaced. I only recently discovered that it's not a character trait of scatterbrained-ness (as I have been led to believe my whole life) but rather high-functioning, off-the-charts, inattentive ADHD. It's why I'm really shitty with numbers, the concept of time, and remembering things. If you dated me, worked with me, or lived with me, you are probably thinking, *Wow, that explains a looooooooooooooot.*

All this is to say that my timeline is reconstructed to the best of my ability with help from friends, receipts, email, digital bread crumbs, and the Doc Martens shoebox of artifacts from under my bed.

Most of all, if there's a moment you recognize in these pages, or a trait that feels as if it could be a bit of you or someone you know, please look through the nesting dolls to understand what influences a person is under, appreciating that we've all fallen prey to the things pulling our strings. Zero judgment from me. Hopefully zero judgment from you, in your new uninfluenced place.

Lastly, I kindly ask that you scribble in the margins of this book.

It is no measure of health to be well adjusted to a profoundly sick society.

—*Commonly attributed to Jiddu Krishnamurti*

Introduction

I was receiving a flower lei at the airport in Honolulu when my life took an unexpected turn. My family and I had just arrived in Hawaii for what I thought was a run-of-the-mill tropical vacation, but my then-husband had a different idea. As my son looked up at me from his stroller, my partner held out his hand and said, "Give me all your devices." This was an intervention. In a sea of Tommy Bahama. In front of a fucking Starbucks.

It changed the direction of my life forever.

For years, I'd been working in marketing and branding, trying to influence people into living certain lives and buying certain things. At the same time I was reaping the benefits of success, I was also experiencing physiological symptoms that I couldn't explain. Not only was I constantly busy and addicted to my technology, but I was also going against my instincts, unsure whether I was on the side of "good." And yet, in opposition to my misgivings, I kept being rewarded. Raises, title changes, and awards were given as positive reinforcement, so I went along with it all, assuming these practices *were* good regardless of how I felt because *so many others said so*—and because to think otherwise would make me an ungrateful, undriven lazybones. (Terms like *burnout* and *quiet quitting* hadn't entered the zeitgeist yet.)

One by one I handed over my laptop, my iPad, my phone. After some serious withdrawal symptoms—itchy brain, anxiety, reaching for an object that was no longer there—I finally surrendered to two weeks without my devices. It was on the eighth day of my forced digital detox that I woke up a changed person. I could smell the sea and salt in the air. I could hear the neighborhood rooster. I could tell you that I hadn't felt this alive, alert, and aware in years.

I found my journal, dusty from being deserted, and wandered onto the lanai. Curling up in the hammock, I began writing for the first time in a long time. What came out was, surprisingly, a list. On the left side of the page, I wrote things like *family, freedom, my time, music, nature, books, creativity, simplicity*. These were my values. Who I was and what was actually important to me. In the neighboring column, I wrote a list of what my current life consisted of: *work, computers, stress, constant overwhelm, sad desk salads, scrolling, meetings, and meetings about meetings*.

Something wasn't adding up, and there it was in plain ink.

I was sick, addicted, and lost.

My path to this point has been circuitous, to say the least. As a young person starting out, I was someone who bucked the norm and resisted the status quo. And yet decades later, I found myself miles from that rebellious lifestyle, instead filling the mold of the stereotypical, plugged-in executive, wearing *busy* like a badge of honor.

Entrepreneurial bosses had me believe we were changing people's lives for the better with their products, so I persuaded customers to buy what we were selling. The corporate feminist manifesto *Lean In* and *#GIRLBOSS* movement not only made me believe I could achieve anything but also that I should do so at any cost—and so I single-mindedly climbed the ladder, hell-bent on success. The tech industry made me believe that exponential growth, speed, convenience, and 24/7 availability were mandatory in the modern world, so I became addicted to their "attention economy." Wellness gurus and the advertising industry made me believe I always needed to be bettering myself, so I walked around feeling incomplete, examining myself for personal qualities I might improve.

I also believed my own hype.

I was told that dance classes helped bolster struggling relationships, so I persuaded retirees to spend a portion of their life savings on lessons, winning me a national "top salesperson" designation. I bought into the pitch that whole foods packaged in plastic grab-and-go packaging were helping families feed their kids something nutritious (never mind polluting the planet), and as the brands sales ballooned from two million to two hundred million in two years, I assisted in positioning a top business magazine to write about the exponential growth and not the exponential waste. After a decade of helping clients find their digital voices and climbing to the top of the career ladder where the stage invites and microphones are handed out, I realized I had helped create a society that was no longer present or connected at all.

I had become someone who signed up for the narrative that corporate culture fed me, influencing others as I clawed for some elusive sense of fulfillment. Even worse, I had been living a life that someone else choreographed under the constraints of a society rotting under late capitalism, that told me what to want, how to be, where to look, and insisted that if I could only become my best self, I could be happy. Suddenly, it occurred to me: *What if I didn't want to be their version of my best self.*

I often return to a claim made by Jungian psychotherapist James Hollis. He basically said that the question to ask when you're faced with a fork in the road is whether a specific action will expand or diminish you. Not "will this make me happier?" but "will it make me bigger?" It helps determine whether you're rushing away from or toward something. Walking away from all the things that society told me I was supposed to want and that I should be applauded for building despite my unwellness would not be easy. But in that moment on the deck, staring down at that handwritten list in black and white, I knew it would be an evolution, not a devolution. The hard choice was the right choice.

What I actually needed was not more to do, but less. I needed to eliminate the stress, the hustle, and screens, not add *more* trendy structures to my to-do list. And so when I returned home, I quit my job during a company-wide Monday-morning meeting about a meeting. In pursuit of a life tuned in instead of screened in, I chucked my laptop and deleted my Twitter feed, bankrupting my email (closing the account with literally more than thirty-five thousand emails in it and never opening it again). Without that involuntary pause on vacation, I don't know when or if I would have had the realization or self-awareness to see *why* I was lost.

My formerly plugged-in, fast-paced, tech-based career gave me a unique perspective on the power of influence from the inside out. I knew that it wasn't just algorithms and ads that were guiding our decisions, worldviews, and self-perceptions, but practically everything in our culture was telling us to live a certain way, stay online, never be satisfied. I decided to do something about it.

I focused my attention on a new endeavor: *Folk Rebellion.*

I thought I had found my purpose. Designed as a kind of crusading platform, the lifestyle brand and magazine encouraged a return to a more balanced analog way of living, questioning cultural norms, and inspiring a much-needed defiance in our modern world. Eventually, I was convinced that I had a mission to better society by waking people up to the dangers of these digital times, spawning a line of apparel and goods featured by companies like Urban Outfitters and publications like *People* magazine. It caught on like wildfire.

Folk Rebellion was heralded everywhere from *Vogue* to *The Observer,* and as the founder, I was tapped for countless podcasts and interviews, positioned as a modern-day Citizen Jane for leading the movement against tech worship and the cult of busy. It was clear that many people were feeling like me—digitally exhausted, burned-out, overwhelmed, unwell—and hoping for another way.

But even after my Hawaiian wake-up call, I hadn't truly learned. The sway of influence is tricky like that. Without understanding the powers of influence, we become trapped in its never-ending cycle, like pushing a Sisyphean boulder uphill.

It turns out that even in an attempt to fight the onslaught of modern-day pressures, you can succumb to them. I hadn't effectively unplugged at all. In my altruistic quest to spread the good word, I ended up living the opposite of my own ethos. I was still under the influence of a culture that made me, well, me. Having persuaded me to achieve at all costs, that I could always do and be better, my mission became bastardized, again taking me in a direction where I did not intend to go. It was a case of wherever you go, there you are. Once again, I was still allowing myself to be influenced.

I had wrongly thought the problem was just the digital world—overuse or misuse. That the only influence I needed to be aware of, and manage, was technology. I thought I'd already made a lifesaving shift, but it turned out that I hadn't truly reached my turning point. It took a pseudo heart attack for me to realize that it wasn't just that one thing—it was everything and everywhere.

In 2018, I collapsed at Logan Airport between back-to-back speaking engagements. I'd assumed that I, of all people, would be able to strike a mindful balance in life. After all, hadn't I already had my epiphany years before? But I'd *once again* been having physiological symptoms that I'd *once again* been ignoring—memory loss, brain fog, confusion, malaise, dissociation, receding creativity and attention— all the same side effects I'd tried to escape from when I gave up my corporate career. This time, though, it wasn't just exhaustion; it was a multitude of things and influences and pressures that brought me right back to the place I had been working to avoid. *Once again* in a goddamned airport. And now I could add a debilitating heart condition to the list. It was now unavoidably clear that I had miscalculated, despite my best intentions, and it was time to come to terms, and come clean, with what I had gotten wrong. The truth is that most of us don't even realize how much we are being influenced, maneuvered, and manipulated throughout each day. I was too busy trying to navigate this complex and chaotic world to see the truth of it all.

And so I stopped. *For real this time.*

I began a self-imposed sabbatical (that's what I called it), during which I stopped outwardly promoting and turned inward to face the mess of it all. I withdrew from the noise of the internet, marketing, and the business of persuasion. It was my decades in branding, sales, and self-promotion, learning the tricks of the trade and honing my skills for persuasion, that afforded me an insider's viewpoint and arsenal of knowledge that ultimately allowed me to detach from it all, finally understand it, and protect me from it.

It's challenging to adjust within a society that runs on systems meant to keep us unaware and unable to opt out. But through harnessing my previous knowledge, trial and error, and staying true to myself, I have *now* found a way to reside beyond the emotional, logistical, and oppressive confines of influence. And I have discovered a happier, healthier, better way of being—for me. On my terms. Not theirs.

Here's a simple experiment I undertook in my early days trying to get a handle on all the influences around me: Try counting up all the advertisements you see, the jingles you hear, the slogans and logos crossing your path. Include requests for you to do things or decisions someone asks you to make. Tally up the favors, the emails, the texts. Make notice of the opinions shared, whether at the playground or in a headline. Add up every time you buy something or pay for something. Acknowledge your forced compliance like stopping at a red light, accepting a website's terms of use, or arriving at school on time. Double the points for things telling you how you should think, behave, or believe. We are living in a time and place where attempts to influence come from everyone and everything—everywhere. The never-ending onslaught had me giving up my count by lunchtime.

Just becoming aware of the number of attempts within a day to control our behaviors or thoughts is enough to begin lifting the veil of influence. Without the awareness, there is no option to opt out.

It isn't always easy, and breaks and reflection and tallies aren't enough. It's about making mindset shifts and actual changes that are realistic and livable from the knowledge gained in those moments.

It's only now, in the absence (as much as possible) of being influenced or being the influencer, that I can see clearly. It took that second game-changing kick in the guts to come to terms with the boxes I got stuck in and, regrettably, created for others.

During my corporate career, I'd sit around glass-enclosed conference tables with other professionals also filled with their own versions of self-importance and set of influences while we agonized for weeks over the right verbiage to "get around" putting a word like *sugar* on "healthy" food packaging, which I rationalized because, ultimately, I thought our goal was still to bring healthier choices to children. Later, I'd discover that the initiative we were to be involved in supporting, a health-focused change to school lunches, was strong-armed by lobbyists and big food. I regret being part of a culture that would create wellness campaigns to subconsciously make you feel unwell, all so you would buy athleisure wear. A culture that keeps mommy influencers on speed dial to counteract bad press. A culture that blends the worlds of entertainment, press, content, and shopping so thoroughly that it blurs the lines, making news entertaining and editorial shoppable, so that you no longer know whether you're reading fact-based articles or consuming ads. And as sorry as I am for playing a part in ushering in the "new wave" of coercive marketing and branding, propelling an obsession with and reliance on the internet and its culture, building consumer funnels disguised as lifestyle communities, and later on, fostering the *business* of wellness, I am grateful to have amassed this awareness, so I can pass it on to you. I am in a unique position to share this information, which makes me obligated to do it.

My world was rife with contradictions. In one aspect, I was quite literally a part of the problem and in another, I was the prey. I'm of the generation that grew up on the fringe of mandatory seatbelt wearing, knowing but not really caring that smoking was bad, living off processed food that was made by corporations that profited while poisoning us. My peers and I are so filled with cognitive dissonance that we can't even agree on a cohesive generational cliché.

Maybe it's not surprising that I've found my way through the obfuscation and into this role. On some level, I knew there had to be a better way.

A pause allows what we are all seeking: the ability to live our lives deliberately. Seeing the motivations of others is as freeing as it is wildly disenchanting.

It allows us to live our lives as we might actually imagine them ourselves. These moments allowed me to become recentered and rediscover life as I truly wanted to live it. It allowed me to remember who I was and hear *my own* voice again.

I decided to stop creating a brand. There is only me now. I dismantled the business side of *Folk Rebellion* and returned to a company of one (or more accurately, not a company at all) to revisit my true desire of living with more creativity, and less stress, in my everyday life. I realized I could spread the word without spreading myself thin.

That's where this book comes in: *Raising Hell, Living Well* offers you a countervoice. Yours. In the following pages, I'm not telling you how to live, but I am encouraging you to really think about how you're living. From my personal experience of being a bad influence to my professional observations as an award-winning corporate strategist, I collected nuggets of knowledge along the way that all seemed to come down to one thing, namely, influence—using it and abusing it. My hope is to shine a light on the chaos of contemporary life by laying bare the troublesome impact of influence in our culture and offering you the freedom of living intentionally without it.

It's taken me forty-three years, and many trials and tribulations, to get here, to finally see the truth of all the influences I was under. Today I may be unlost. Untrickable. I am aware of the toxic powers attempting to control or maneuver my life, my decisions, my purchases, how I raise my kids, what I consume, how I take my coffee, and even my core belief system. I've studied, mastered, and wielded influence and still fallen subject to it. I am sure there are more lessons to learn.

I now realize that my path toward living free from influence won't ever be complete but rather will be an ongoing adventure. And learning that the path won't be a simple one was necessary. It helped to put all the puzzle pieces together, to recognize that I held the knowledge to wake people from their present-day slumbers with the hope that they'll rise up against our society's pervasive influences and its effects on us, so we can all—me, my family, you and yours—be free.

Raising Hell,
Living Well

Sometimes in life you should stick to your worldview and defend it against criticism. But sometimes the world is genuinely different than it was before. At those moments the crucial skills are the ones nobody teaches you: how to reorganize your mind, how to see with new eyes.

—David Brooks

Part I
Under the Influence

There's a riddle you might have heard before. The first time I heard it was in the third grade in 1991. I don't even remember the name of my teacher that year, but the riddle stayed with me so much that I later used it to open speeches, in workshops, and when hosting events. It goes like this:

A father and son are in a horrible car crash that kills the dad. The son is rushed to the hospital. Just as he's about to go under the knife, the surgeon says, "I can't operate—that boy is my son!" Explain.

Your answer says a lot about who you are; or more accurately, it says a lot about what made you who you are—elements like which generation you grew up in, where you were raised, and your education and experiences. It's not just your personal experiences that affect you, though. You are who you are today based on the evolution of everything that's happened throughout time, leading to this very moment. You, and the rest of humanity, are shaped by the choices and decisions of people and things that came before.

This is called influence.

That's why it was no surprise to me that the singular person in a room full of professionals to solve the riddle was a little boy, brought by his mother to a panel on the topic of "Women in Wellness."

When he couldn't take the "Gay dad!" "Stepdad!" "Grandfather!" "He's still alive!" or "It was a dream!" answers anymore, he saved the other audience members from themselves by shouting, "It's the mom!" A collective groan filled the room as almost every woman in attendance shrank in her seat.

You might be wondering the same thing as all the women at the wellness event: How did my mind not see such an obvious answer? What made imagining a surgeon mom so difficult?

Influence is incredibly powerful.

It has the power to create your beliefs and your values, impact how you think and perceive the world, and alter your behaviors. The kicker is this happens most often without your ever realizing it.

Movies like *The Truman Show, Free Guy,* and *The Matrix* depict these sorts of wake-up moments where the main character realizes there is a world outside their own perceptions. The more self-aware the character becomes (of themselves and the world in which they live) the more free will they actually have.

I'd argue that these movies were influenced by one of the oldest philosophical ideas about influence: Plato's *The Allegory of the Cave*. The allegory begins with prisoners who have lived their entire lives chained to and facing a wall inside a cave. Behind the prisoners is a fire and people they can't see, who are carrying puppets in the shape of people, animals, and objects. The puppetry creates shadows, which are projected onto the wall in front of the prisoners. The captives observe these shadows as real things, not puppets, forming their reality because they have no other experience.

The first step in living outside the metaphorical cave, the matrix, or the artificial reality is acknowledging that everything we know is based on a collective paradigm. Shared rules, assumptions, and ideas contribute to our worldviews, which is something we have gradually adopted as true and accepted. This is the foundation for the framework from which we operate every day. Upon leaving the cave, some people become so blinded by the light that they would rather retreat than see their beliefs challenged. The illusions we are under have shaped how we think, what we say, and how we do things, but when the light seeps in, what was once considered normal and natural suddenly appears darker.

There are myths (shadows) that we grow up with that are so commonplace in our minds that we accept them as fact instead of opinion. We breathe them in on a daily basis until we assume they're the norm and adapt our behaviors to meet the demands of the accepted environment—our cave.

Accepted myths come in many variations. They can run the gamut from concepts like men should be strong and women should be pretty to the idea that marriage or buying a house signifies adulthood.

For the purpose of clarity and ease, throughout this book I will be calling these commonplace myths *Folklore*.

folk·lore /'fōklôr/ *Folklore* is the expressive body of culture shared by a particular group of people; it encompasses the traditions common to that culture, subculture, or group.

Folklore are the stories told to us.

Generally speaking, people want to belong. To belong to their community, a person often unknowingly conforms to the beliefs and expectations of the group. They may not realize they are seeking approval, but their fear of suffering disapproval is strong enough for them to contort themselves to fit within the *Folklore*.

We've all done it. We tell ourselves things that allow us to remain unnoticed among the herd. These things are often made up of semi-untruths, ignorance, a lack of self-awareness, or outright lies. Let's call them *Folktales*.

folk·tale /'fōktāl/ *Folktale* is a tale or story that is part of the oral tradition of a people or a place passed on by word of mouth.

Folktales are the stories we tell ourselves.

I have to go to college. Having this bad relationship is better than having none. I must maintain my youthful appearance. I am happy climbing the corporate ladder. I could never move and uproot my life. Buying this will make me feel better.

The *Folktale* feeds the *Folklore*. The *Folklore* needs the *Folktales*.

I'd like to add another option, the *Folk Rebellion*. What if for every tale and lore there is a moment in which you pause, assess, and see the influences for what they are—leaving you the option to make informed, intentional choices yourself. You are not tricked by shadows. You are not automated. You are not trodding a path worn bare by the followers of the tales and lore. You are not telling yourself lies, or living half-truths, to conform to the "that's just the way it is" attitude of the world.

The *Folk Rebellion* is the antidote.

It's your awareness and intention set into action against the influences of your life. The Rebellion is what we will be seeking for each step of the way throughout this book.

For example, my experience in Hawaii might look like this when considered through this framework:

Folklore   The world is undeniably digital, and you better hop on or you'll be left behind.

Folktale   I must be connected 24/7, adopt all new digital advancements, and champion all innovations to be a part of society.

Folk Rebellion   Technology should work for me, not the other way around. I can choose to use it as a tool and not a replacement for real life. Every sparkly new thing should be weighed against my values and what it's replacing if I decide to introduce it into my life.

Pretty freeing, right?

To be free from the crushing, mad-making pressures, it's essential to understand where these powers that are trying to influence you come from, why they're doing it, and what you can do to manage them. That's how you can avoid having *them* dictate *your* decisions. We'll cover all that (and so much more) over the course of the book, but in Part I we'll start with the "where." Where are all these influences coming from and how do we spot them?

In short, influence lives everywhere. It's in the very fabric of our society, from our media outlets to the way our workplaces are structured, in the way our families are formed, in the way we keep time, and in the way we drink our coffee. It's embedded in the compromises we make with our partners, guiding how we raise our children, the way we talk politics with friends and family members and strangers on the internet.

Let's break this down a weensy bit more.

There are many sources of influence making us who we are. Of course it's hard to know who we are when we can't hear our own voices. To help illustrate this as we begin our exploration and reclamation, I've created these spheres as archetypes to guide your understanding:

Spheres of Influence:

   1. *Inner World*—The place below the surface filled with the things that make us us. Fixed things like *biology, genetics, childhood,*

experiences, age, family lineage, birth order, personality traits, and changeable things like *knowledge, attitude, behaviors, skills, hobbies, beliefs, values, expectations, characteristics, health, needs.*

2. **Surface World**—The real-world reciprocal relationships around us, the surface represents where, and with whom, we are present and in community. *Family, friends, partners, co-workers, school, neighborhood, gym, participation sports, places of worship, local business.*

3. **Outer World**—It is the universal atmosphere of the collective omnipresence of intangibles floating all around us. *Business, brands, media, tech, norms, culture, the arts, trends, all the isms, beliefs, zeitgeist/trends, social media, content, books, movies, celebrities* and also *established structures and institutions, infrastructure, resources, policies, algorithms, politics, leaders, regulations, religion, education, government, history, socioeconomics, evolution, environmental conditions.*

Today's world is noisy. It is coming at us 24/7, and no place is safe. That's because all parts of our *Inner, Surface,* and *Outer Worlds* of influence commingle and affect one another as they come at us from every angle.

Your doorbell, your cell phone, your Amazon account, your streaming music, your algorithms, your maps, your apps, and even your well-intentioned loved ones analyze you and tell you what to do now. We are more desperate than ever to figure out how to unhook from the things making us feel sick, tired, lost, stressed, angry, and overwhelmed. We are not ourselves when buried underneath all of this influence.

And it can be a confusing paradigm to navigate! We are constantly being told we have all the opportunity we could ever hope for, all the answers within a keystroke, and all the comforts and conveniences of the modern world right at our doorstep. And in some ways, maybe we do. Yet more than ever, the general population is very unhappy and unwell. From morning to night, we're being assaulted with marketing and advertising jargon so ingrained in our culture that we don't even realize that we've become its parrots, too, repeating these ideas to our own spouses, kids, friends, and co-workers. We argue that "the early bird gets the worm" as we fight anxiously online to get tickets for some must-see movie or play. We live the tenets of "You snooze, you lose!" as we battle to get our kids into the one preschool that matters or work day and night to get a promotion over someone else. We spend hours perusing pictures of friends on idyllic vacations online, feeling terrible about our own lives, only to mimic that behavior and do the same ourselves, presenting only our most perfect selves. We further the toxic normality (including normalizing the use of the word *toxic*, therefore stripping it of any true meaning!), while yearning for validation. We buy into capitalist convenience even though it's come to replace the legitimate human connection that brings us fulfillment, leaving us lonelier and hungrier for more quick fixes from fast food to fast fashion to fast fucks. When everything becomes a shortcut clickable commodity, the clicker—who ordered a burger or a person to their front door—is left feeling gluttonous and guilty, but still unsatisfied. When we look through the layers of influence to see who benefits from what, we can begin to see beyond the shadows into how we are not doing our bodies, our minds, our lives any "good," but, in fact, are doing each of them a disservice. It can feel overwhelming and disorienting to find your way through such a dense forest of influence—because it is.

Times may have changed, but the concerns about influence have not. It's so powerful and pervasive, so ingrained in our being, that even the world's ancient philosophers spent time contemplating it. Identifying the influence is always the key step. It's the hook in the parable. The great *ah-ha!* in our modern stories and ancient philosophy. It's the seeing that is the pivotal moment—the waking up, the unmasking—the turning around to see the puppeteers behind you. Once we begin to recognize the many influences impacting our worldviews and behaviors, we can begin to crawl out from under their sway. And until we venture outside the confines of our own caves, how we think, what we say, and the perspectives we hold on to will be based on someone else's projections.

Today, it's how a roomful of women advocating for gender equality who grew up under twentieth-century worldviews—that women weren't equal to men, perpetuated through everything from commercials to TV shows rife with gender stereotypes—still can't imagine anything but a male doctor in their mind. Change happens slowly, but for a child who comes of age with a female U.S. vice president and female pediatricians, a mother as a doctor is obvious.

Influence is everything.

Moving beyond the scope of influence is so important. Without it, we end up wasting our years living by the rules of societal norms created by corporations, brands, politicians, start-up founders, and megalomaniacal tech moguls. It affects everything from our self-esteem and finances to our independence and risk assessment. Without the powers that be wielding their influence over us, we can live fuller lives more freely, with *our* versions of "well-being" and "a life well lived" taking center stage.

And then, in the best-case scenario, we can begin to use our own influence for good.

So without further ado, let's get a move on and begin to uncover where we find all this influence in the first place!

TRACKS

(or Our Inner World: Influence Below the Surface, the Direction It's Pointing Us, & Aspiring to Cross It)

Upstate New York is the bologna sandwich of New York State.

Utica, New York, specifically, is something of a caricature—the satirical target of American cultural cornerstones from *The Office* to *The Simpsons,* in which Homer famously said that the area could "never decline because it was never that great." It's a late show punch line, roasted by hosts. It's objectively outdated in its architecture, infrastructure, and industry, but also at times, in its beliefs and worldview. It's like that uncle who's just a little bit inappropriate, but he's from another time and mostly means well, so you try to let it slide and love him anyway.

Some would call it a joke. I call it home.

Having been raised in the heart of that Wonder Bread sandwich, I am forever both defending and judging my roots. But in reality, it's that polarization that created the foundation of who I am and how I approach the world. I am that tension. After all, the influence of where we begin can determine a lot about where we end up.

Borders, zip codes, tracks. Wrong side. Right side. These are among our very first influences that blend beneath the surface of "us." Let me show you what I mean.

When the railway was first invented, neighborhoods near or downwind of the tracks bore the brunt of noise and soot from the locomotives. Naturally, those areas quickly gained a reputation for being socially inferior. Literally and figuratively, the tracks divided the more prosperous from the poorer. They separated those with good fortune from those who drew short straws. Ever since, tracks have represented a not so invisible line. They are the straight of an arrow pointing in the wrong or right direction, depending on which side you are on, and a reminder of a possible way out if you could only follow the wooden planks beyond their known world. As a little kid, I crossed over the tracks on yellow school bus number 223. When that bus bumped over those tracks, my three siblings and I would lift our feet, cross our fingers, and hold our collective breath, making wishes for good luck. For me, the tracks were everything. They represented the ways in which I didn't have enough and the ways in which I had more than many. They were my escape route and—to this day—my path home to myself.

Utica itself exists as a kind of "other side of the tracks" to its more well-known and well-loved counterpart, New York City. Or at least that's how the people who live in Utica feel. In reality, most New Yorkers have barely heard of it. The competition is one-sided, an inferiority complex well earned.

It's also a mishmash of commingling opposites from opposing sides of the tracks—old and new, rich and poor, Black and white, educated and uneducated, red and blue. Our small footprint is home to a whopping seven universities and colleges, but also faces an intense brain drain when so many of our educated young adults flee after graduation. We have ostentatious McMansions built on tall bucolic hills overlooking valleys with jails on the other side of them. We have a diverse mix of ethnicities and races, but segregated towns. We have centrists and apolitical types—not to be confused with the liberal hippie havens of downstate. And moderate is about as progressive as the majority of people get, living next door to secretly staunch Reagan-era Republicans and die-hard Trumpers. And it's been a place throughout history for immigrants to start over. But when Utica more recently opened its arms to refugees—in the '90s to Bosnians following the Bosnian War, in the early aughts to Burmese Buddhist monks after the Saffron Revolution, and, as recently as 2021, to Afghans after the fall of Kabul—it was often the second- and third-generation descendants of immigrants who resisted "those people" coming to "their city" to "steal" their jobs. It was as if they'd forgotten that their families were once from somewhere else too.

I was made there, born to an Irish mother from the "right side" and an Italian father from the "wrong side." So maybe it makes sense that even my immediate family was rife with contradictions: I was raised to believe there was never enough and always enough at the same time. I was taught to respect authority while also to challenge it. I was sent to a religious school by an at-the-time agnostic father and a now-repentant Catholic school girl who never lost her faith in God but did in the church. While conservatism was historically our family's brand, the beliefs of the household were mostly liberal. I was a "free lunch" kid at the local suburban rich school, hiding my dime stipend below the milk jug. And maybe because of all of that, I was the first kid in the family to gain admission to a fancy liberal arts college—and the first to toss it aside in search of something more. Influence is wily that way.

My parents—who divorced when I was thirteen—will each tell you that their side of my heritage is the better half. Family mythology, spun into lore around our kitchen table, told us that the mix of our genetics—Irish and Italian blood—made for beautiful children with bad tempers and, later, drinking problems. It's this sort of objective bravado that defines the people from my hometown—slightly self-aware, deeply self-deprecating, and unbelievably stubborn. At once, celebratory and dejected. In my experience, that cocktail makes for really fun, warm people with an edge, great conversationalists full of spirited ribbing and quick wit—but only if they like you. The people have big hearts but also hard noses and chipped shoulders. And it's one of the reasons I think I've tried, and failed, to live happily in many other places. When sarcasm and grit is such a part of your inner makeup—a love language of sorts—those characteristics make it difficult to connect with people who would happily trade irony for a balmy 72-degree, sunny existence. That's just not the Utican way.

Known as the original Sin City thanks to its history of organized crime and political corruption, my family lineage reflects that in spades. Putting Utica on the proverbial map was a great, great uncle from my mother's side who was the vice president of the United States under Taft. But it was a more questionable relation—a distant great-uncle on my father's side who was the proverbial backroom political boss of Utica—who took it off. I have experiences of baby grand pianos, afternoon patio cocktails, three-piece suits with bow ties, and Cadillacs. And memories of backyard tin can tomato plants and plastic-covered sofas, dingy bars, old cars, and even older card sharks. Just the right amount of high/low to help me develop "character."

My hometown is chock-full of upbringings that molded beautiful humans, but also created odd friction. People love it and love to hate on it. Utica is equal parts charming and befuddling hot mess.

At times, you could describe me that way too. It's not surprising. After all, our origin is one of the greatest influences on us—where our ancestors hail from, where we are raised, where we choose to live, and even where we fall within our family order—origin sets us on a track from which it is hard to veer. Place shapes us. The influence of these chance factors, some fixed and some not so much, is incredibly formative, affecting our *Inner World* from a young age. For the lucky, this means being born with a horseshoe up one's ass; for the ill-fated, it's bad trot.

The first time I got drunk was on some still-functioning but old railroad tracks. They were situated next to the local strip mall at the bottom of McMansion hill. Next door was the police precinct, which should have made it a dicey spot for underage alcohol consumption. And yet, there we sat. In the glow from the parking lot's single light, my best friend Roxanne and I could see each other's shadowed thirteen-year-old faces, my apprehensive expression reflected in her Coke-bottle glasses. At the time, I was a somewhat naive rule follower because I hated to get in trouble and had no older siblings to show me the ropes. Also, my parents' child-rearing philosophy for me, their firstborn, could be summed up as high expectations for everything, low tolerance for bullshit. As is often the case with parenting strategies, this philosophy waned as the years passed and the number of children grew. For my three younger siblings, tolerance for bullshit went way up. I'm still a little bitter about the long-lasting, somewhat unshakable influence of our birth orders. But in the meantime, my friend Roxanne was my "bad influence," an only child who never really got in trouble and whose father, Tony, doled out twenty-dollar bills from a rubber-banded wad of cash. We were just entering teendom. Yet in the stones and dirt on the railroad ties, Roxanne and I sat, mixing room-temperature vodka stolen from my pop's liquor closet with too little gas station OJ. Things were headed in a poor direction.

Being on the tracks, as Roxanne and I were on that night, was in some ways preferable to being on either side of them. During my childhood, I had a clear sense of existing just barely on the right side.

Mine was a childhood full of incredible freedom, of presence, of folly most every day. And the more I pushed the boundaries of the town, the more brazen I got. The more rules I broke without repercussions as I got older, the less I felt they applied to me. Lots of high jinks. Zero fears. Just enough bullshit and privilege to create the inner foundation for whom I would later become. Someone who pushed boundaries in life. Someone who took risks because I knew I always had a safe place to return to. Someone who was supported, loved, independent, but full of piss and vinegar with something to prove.

I realize now that in many ways, it was the picture of childhood in the nineties, a time before the internet stole our attention, anonymity, and spontaneity. Before our folded jotted notes were replaced by text messages and our mixtape love letters with swiping. Before everyone knew everyone's location, what they were doing, and with whom—in real time. It was the last era when, as a kid, you could get lost and in trouble without having a personal brand that you couldn't escape because it lived in perpetuity online. I could choose to be a fuckup one day and a saint the next.

I don't say all of this to wax nostalgic. But the deeply ingrained influences within us don't stop at location. Where we are born, where we fall within our family birth order, where our ancestors hail from, are incredibly powerful in the grand scheme of our lives, but we can't look to them without looking at each of them within the context of *when*—a place, a family, a history, even our health changes with the times in which they exist. If *where* is the map, *when* is the legend we use to read it.

The nineties were not perfect, but they were my *when* which informed my *where*. The already hugely significant and unavoidable influences of *when* and *where* we lived are overlaid with the even more influential digital revolution for anyone born of the Gen X nexus of before internet/after internet. This key fact—that I was born of this time—informed so much of who I became.

And like a lot of "back then," my youth was both idyllic and problematic. My internet-free formative years mirrored what was depicted in movies like *Stand by Me, The Sandlot, The Goonies*—a bunch of kids fumbling and figuring it out. Sheltered from the world outside our place in it and drunk on the freedoms of all that our origins provided us.

The fact that I wasn't concerned about getting my first buzz on so near the police station is also an indicator that not all zip codes are created equal. Looking back, the location Roxanne and I chose for our first drink seems odd. It suggests a sense of invincibility, no fear of consequences, if there were any. The truth is that that sense of invulnerability didn't stop there, a dangerous mix of youthful high jinks and entitlement. Hell, my siblings and I threw notorious rager parties that were shut down by cops who'd kindly leave behind twelve-packs for us to imbibe while we cleaned up the mess. I didn't see at the time the entitlements we were granted. We weren't taught about things like privilege at our privileged school, even if I lingered at the outskirts.

When our zip code wasn't enough, the luck of winding up in a close-knit family was plenty. Mom would say we were rich in love. Our home was one of warmth, safety, and laughter. We may have been on the low end of the high-end town, but we never lacked for anything; all our needs were met, and the most important one was love.

And that upbringing laid the groundwork for what would become a complicated but also pretty darn charmed adulthood for me too. Don't get me wrong—I've had plenty of trials and tribulations, but I learned certain values that have stayed with me regardless of where I've gone in life since leaving Utica. Being downright forthright was almost a birthright where I was from. Neither of my grandmothers minced words. That skill to say what you mean and mean what you say is something I am proud to have taken with me. My grit, often

acknowledged by others to this day, can be traced back to the spirit
and expectations of those I grew up around. Tenacity and persever-
ance were what brought immigrants to Utica. Having just enough necessity and not enough support created a communal hardiness that
runs like warm lifeblood through the area, and now my veins and the
communities I belong to too. My upstate roots have made me *feel*
lucky even when I wasn't. Understanding the belief that "life isn't
fair," a common refrain uttered throughout my childhood, seems to
have given me the huge gift of easier acceptance when things don't go
my way. Thanks to my mother, I was taught to value sentimental
things like books that were passed down, borrowed T-shirts, and family recipes more than designer labels, fancy cars, or money. To this day,
my most prized possessions have no economic value, but they are truly
invaluable to me. Foolhardy guts, an attitude and behavior that has
served me well, came by way of sibling double-dog dares, blue collar
barroom banter, and the fact that when you grow up thinking you
don't have much to lose, your mind leaps to *What's the worst that can
happen?* And most important, I learned (many times over) that we are
as strong as our community. My ability to live freely out in the world,
from misadventure to misadventure, sprang from the knowledge that
no matter what, I would always be okay because, as my family, friends,
and former teachers would say, I could always "come home" if things
went awry. What a gift.

And while some of our *Inner World* influences are fixed (try as you
might, you can't change your birth order, your blood, or your
childhood—sorry!), fortunately, there are aspects of our *Inner World*
that can be remade time and time again. For all the love I have for the
things Utica gave me, I'd be remiss not to mention that there were
some not-so-positive influences that also made their mark and that
I've continually reflected on and worked to overcome even to this day.
For instance, my foundational knowledge, and therefore my initial
views, came from my school, where most of the children were white,
and all of the teachers were white and followed a curriculum to match
those systemic influential issues. When I left home, both on purpose
and by happy accident, I expanded my knowledge of a world outside
of the only one I had known. By doing that through living elsewhere,

expanding my relationships, and actively seeking out opportunities to learn about a wider scope of worldviews, my attitudes about people who looked different from me, had different fixed inner influences, or lived differently in different places, became, well, different.

Not surprisingly, the more we enter the *Surface World* and *Outer World* of influence outside of ourselves, the more we can change—for better or worse, depending on our choices.

As a kid, I assumed the world everywhere looked like my world. And as a child, I believed that anyone who lived differently *chose* their fates, not that they weren't granted or given the same opportunity as I was. I couldn't see past my own nose, my zip code, my house on the hill at the "right" school on the archetypal right side of the tracks to understand that my carefree youth wasn't available to everyone.

But from the vantage point of my current Brooklyn zip code, not too far down the tracks, I can see it clear as day. I can see how where I sit today in essence began where I sat on a map the day I was born. I can see the influence my home had on me, for better and worse, and I can see how it's possible to both hold it close and escape it. My present day might look far different from my younger days, but the *Inner World* values and beliefs are still the same. In fact, it's those very things that led me to this location and this point in my life! And for that I will be forever grateful. Tracing values, beliefs, or behaviors back allows us to take stock of where, how, and why they originated, and to see how they've morphed over the years. In doing this, we can determine what's serving us and choose to hold it tight, or what's harming us and finally let that shit go.

Without the origins that created my take-it-on-the-chin, one-day-at-a-time mentality, I might've become one of those people who lived in New York City one time but found it hard and left, never to return. Oh, the horror! Instead, I've lived in the city too many times to count, having left and fought to come back each and every time. So thank you, Utica, and specifically that teacher who said, "Life is tough—get tougher." I've now lived in New York City longer than I've lived upstate, and yet both are home.

The sandwich I'm in now might be more artisanal pastrami on rye than bologna, but I like to think I also keep some of that great upstate flavor here with me.

Folklore   Where you're born dictates who you are and your future.

Folktale   I must conform to, accept, and live within the identities, worldviews, and borders of my environment.

Folk Rebellion   I can take what's valuable from my origins and change direction, switch, or cross tracks by choosing to live outside the influences of my youth and its surroundings.

Raise Hell   Take the time to think about your values and trace their origins. Ask yourself if these values are ones that still work for you today, or if they were accepted because that's what you were first introduced to. Looking at your *Inner World*, what is fixed and how did it make you who you are? Now, what is more malleable? What would you want to modernize, expand, revisit, or change?

PICKET FENCES

(OP Our Surface World: the Influence Next Door, Expectations of Settings & People, & Attempting to Pause)

When I was seventeen years old, I wrote a book called *Fuck the White Picket Fence.*

Or to be more exact, I scrawled a loose outline and a few chapters of a book reflecting my ideals as a teenager who was feeling backed into a corner. When I thought about my future, I wanted to be successful—I knew that, for sure—but I also wanted to define what that meant to me and determine my own path for getting there. I felt resistant to the standard, unwavering road map that people subscribed to in the community where I was raised: college, career, marriage, house, family.

Say what you will about teenagers' underdeveloped prefrontal cortexes, to this day, even as someone who has achieved a degree of that societal success, I think I was onto something.

I hated that I was expected to define my future through neat and tidy answers to commonplace perfunctory questions from my teachers, family, and friends. "What do you want to be when you grow up?" was already ingraining the idea that we children should have established desires, goals, and a full sense of identity somewhere between fifth-grade recess and puberty. I was sensible enough to realize that this influential sense of expectations—which began at a very young age—was bullshit. I couldn't make truly informed decisions before I had really lived! I also resented being compared with my peers, some of whom had already chosen tracks or personal brands for themselves—dancer, firefighter, veterinarian, pediatrician. Easy identifiers with clear steps to achievement. On one hand, I felt suffocated by the idea of mapping out my life that way, without room for surprises or discoveries. On the other, I was envious that they knew—and I didn't.

I hated that the elements outside my front door, the settings and relationships of those closest to me, were attempting to cast me toward the direction that paved the way down that road, a blueprint that began with dreaded early morning wakeups for school that didn't make sense to my individual self then and *still* don't. And once there, I struggled against all the memorization, the competition among peers, and the expectations that felt thrust onto me by the authority figures and leaders around me.

In all of this, we find our next major sphere of influence: social expectations and situational standards forged from the relationships of the communities immediately around us. It's our close reciprocal relationships—the two-way streets—the influence of who and what are near to us.

Back then, some might've called me a tad bit rebellious. The truth is that not everyone, me included, fits into the one box that society and its structures predetermine for us, especially the ones formed in our more immediate surroundings. I found school and all its constructs boring, unchallenging, and less inspiring than a good book. So I hid behind manners, smiles, and good grades while simultaneously cutting class and corners, earning myself the nickname Sloane, after Ferris Bueller's troublemaking girlfriend.

Even with the tomfoolery, I was able to keep my grades up and, to the shock of my principal and some townspeople who expected someone with my pedigree to have less lofty goals, I was accepted by the college of my choice on an academic scholarship. No one questioned whether I'd go. This was the nineties, the time when more schooling meant more success, before tech tycoons in hoodies eschewed college and made it cool by making billions in doing so. There was an entire *Outer World* system designed to get me to go, a pipeline that begins in our *Surface World* before we learn to walk. After all, even the wrong nursery school can set you off on the supposed wrong track.

College was a setting that positioned my education in terms of its benefits to me and the career I could trade it for once I'd completed four years. But the experience is one that is also under the influence of a growing system that prioritizes maximizing revenue. It's given rise to an increase in for-profit colleges, banks doling out high-interest student loans, expensive admissions coaches, and various consultants— all to ensure that I would take the *right* (profitable) steps toward success. It's capitalism . . . sold as *promise*.

Meanwhile, the other nonoption—skipping college—evoked local fear and scorn. And so I strategically applied early decision to the college of my choice, knowing that would increase the likelihood of acceptance. The deeply ingrained higher education complex even had me falling for one brand of college over the other. It wasn't really about what I would learn, what job I might be guaranteed afterward, what the net was on debt-to-earnings-potential ratio. What it came down to, in reality, was the bumper sticker—an emblem my parents and I could place on our cars in our hometown, a logo that would be proudly worn across our chests on hoodies and T-shirts to the gym, the office, and in front of the neighbors. A symbol of status. The college I chose was one of those schools that many in my town wanted to go to, and I knew earning a spot there would make me seem better. Better than what, I wasn't sure. But it felt important. Like a chance to shout, "Look at me. I'm getting out of here!" The road to a bright future was clearly mapped out for me, and now, I had the college swag (my outward visible value to showcase and influence my *Surface World*'s perception of me) to prove it.

Of course even a "free ride" comes with fees (another situational standard we're supposed to accept without question, right?). Once I was admitted to college, I bought right into that community's accepted way of thinking and began slogging down the path so many others had taken before me.

I worked multiple jobs for cash I hid in a shoebox to pay for my dorm room, meal plan, and books. When I wasn't selling snowboards and mountain bikes for the local sports store, I was a yogurt slinger at the mall. I rolled cigars and built humidors for an eccentric entrepreneur, who controversially paid undiscriminating high school students $20 cash per hour when minimum wage was $4.25 and let us smoke cigarettes in his house in the woods. No brainer. I also babysat and played counselor at the local rec center, ultimately amassing just over $7,000.

Truth be told, I was still a little short, but I was confident that once on campus, I could buy some used textbooks to make up the difference. I was already this far down the road—surely the rest would fall into place as my new community made it seem it should.

You see, once you decide to operate within a group or community's given system, that system's pre-existing structure puts you on the conveyor belt that keeps the system running. Sometimes that's okay, and sometimes it's not. What's most important is understanding, and then recognizing, that there is even a system to begin with! Choosing whether we want to be a part of it is easier once we know it's there, how it functions, and why. Often what keeps these deep-rooted systems running is our adherence to the structures that benefit the system, and not always necessarily what's best for us.

Education is a structure within the overall system, and of course it can be (and should be) one that is beneficial to all. But if we were to look at the way the system is designed, or what this structure has morphed into from its origins within the system, we might see that it's another business, or another brand, or another opportunity whose value and accessibility changes for us depending on locations or socioeconomics or background. We should see that our personal mix of *Inner, Surface,* and *Outer World* influences determines how we experience it. We'd be able to recognize that debt in exchange for a degree may not make sense for everyone. Sure, yes, there are professions that absolutely require degrees, but my peers' accepted standard that the only way to become meaningfully employable is within the confines of the pay-for-play educational establishment is complete and total horseshit. It's horseshit *and* a financial model that positively affects banks, capitalism, and future companies that just want to hire someone with a big degree and bigger debt, thus creating a good employee who's tied to the system by having to work hard, fall in line, and achieve more to pay it off.

Here we can see that while the source of the influence starts with a communal norm, if we dig deeper, we see that the influential forces at play in the *Outer World* perpetuating this norm for their own gain are manifold. This is an example of one structure within the system that attempts to, and succeeds, at keeping us on the conveyor belt, therefore making us dependent on that very system. We'll explore the *Outer World* more thoroughly in the next essay, but for now, just know that our *Surface World* is composed of influences like community expectations that are in turn *also* highly influenced by *even bigger* outside sources.

I used to frustrate my parents when they emphatically stated a "must," and I would respond, "And then what?" over and over again like a child playing the never-ending game of "why." You must get good grades. You must be on time. You must go to college. *You must, you must, you must.* To which I asked, *And then what, and then what, and then what?* It was exhausting but usually proved my point. When you look at the whole system, and see the "must" that's the first stop on the local conveyor belt, asking "And then what?" until there is no answer left allows the real sources and spheres of accepted influences, like norms and systems, to reveal themselves. At that point, you (or an eighteen-year-old me) could say, "But I don't want that life" if, upon further reflection, you do not.

Though I wasn't sure what I wanted to be when I grew up, and fortunately hadn't pigeonholed myself, I had some clear ideas of what I didn't want, which felt like a start. I also had a sense of what interested me: I liked stories. Writing and making things has always been a natural part of my life, something fostered and recognized by those closest to me. So when I looked at the course catalog, communications and media seemed like the natural fit. It wasn't until orientation that I began to understand exactly what I had signed up for.

Seven thousand dollars was a lot of money for someone who was scrounging for minimal hourly wages. Each dollar represented a missed night with my friends or one less concert ticket. So when I arrived on campus and went to register for relevant classes, only to be told that not even one was available, my seven thousand wishes seemed to disappear down a well. This was shaping up to be a blow-off semester, something that I imagined kids who didn't pay for their educations might construct on purpose: a semester filled with underwater basket weaving and fuck-off rocks for jocks' classes. But I didn't want to coast. I wanted to squeeze the opportunity out of every red cent I had saved for this momentous make-or-break rite of passage. In my eyes, a sleeper semester was equivalent to setting my shoebox full of money on fire.

The communal standards become so ingrained that people everywhere mindlessly set shoeboxes of money on fire throughout their lives. It's hard to stop and question whether what you are doing is the right thing for you, especially when those around you accept these practices without a moment's pause. If everyone is taking loans they can't pay back and buying textbooks that cost more than a vacation, then taking your shoebox and pausing before you set it on fire makes you the odd man out. Your pause, and the questions and choices that follow, can feel like a threat to those who have already lit the match.

After orientation, I returned home, driving the whole whopping two hours between the campus where I was going to spread my wings and the home I had known for eighteen years. On the way, it occurred to me that the college I'd chosen wasn't that far away nor was its location all that different from where I'd been raised. If one of my primary dreams was to see the world, why had I chosen a school essentially down the road?

Thoughts like those began to creep into my psyche in the late summer of 1997. Maybe I was trying to poke holes in my plan because I didn't want to waste $7,000 and didn't know where I'd get the *next* $7,000 for the following semester's books, bed, and food. Or maybe it was because deep down I felt like I didn't deserve this rarefied education. Maybe I was afraid of failure, and self-sabotage was the only way out. Maybe it was the fear of losing my first love, who was planning an adventure out west, without me. Whatever the impetus, I couldn't ignore this icky feeling that was slithering in from my lizard brain. And that's what pushed me to pause to widen my lens, stepping to the side of the conveyor belt for a moment before stepping on. I wanted to be sure my actions reflected my desires, and that the structured path I was about to embark on would keep me empowered, not indentured.

The pause was during the last days of that August. I stopped and looked around before taking the next step down the expected road to success. My little town in Upstate New York is known as the gateway to the Adirondacks, a place that was and still is one of my favorite on earth. I've always felt lucky to have that respite within such a short drive. While other teenagers in other places were hanging out in malls (for fun—not scooping yogurt), I was hopping into my boyfriend's Volkswagen to get to a place to hike before dusk. Those woods were my stomping ground and, I later realized, where I did my best thinking.

Holed up beneath the stars and the trees, wrapped in a red fleece blanket needlepointed with my name and future collegiate logo on it, I sat by a roaring campfire—far away from the voices and opinions of others—and contemplated my future.

Nature has always been a cure-all for me. My time alone in the fresh air allowed me to step back from all the "shoulds" that were clogging my brain and make room for stuff like analysis, intuition, and big dreams. My pause allowed me to get out from under the expectations of my family, my friends, my community. My pause allowed me to stop and see my trajectory clearly. My pause allowed me to escape the minutiae of my day-to-day routine and take a hard look at my reality.

It didn't take long at all to gain clarity.

In pausing, I was able to ask myself the most salient question: Was I signing up for four more years of school because I really wanted to or because of silent but powerful outside influences? Was I doing this because it was expected of me, and I couldn't imagine another option? Because it seemed "normal" and like what everyone else was doing? Or worst of all, was it because like others I'd grown up within our mutual community, I wanted status or bragging rights—a desire that I was convinced was a typical reward to want?

Those questions in the woods gave rise to others—even bigger and more fundamental: *What do I want? What don't I want? What does success look like to me? What is life really like anywhere besides Upstate NY? And if I want to know the answer, why am I staying here?*

And that's when I realized that my college plan equaled everything I hated about the road to success.

So instead on the same day I was to leave for the first day of the rest of my life at college, my boyfriend and I packed his Jetta and hit the road. With $7,000 in my metaphorical pocket, I dropped a Dear John letter to the dean of admissions into the mailbox and left town.

I wasn't the first person ever to take a leap like this, though it certainly rocked me to my core; a rebellious teenager, bagging college, taking off—with her first love—on a Kerouac-style road trip, hoping to find herself. It's a trope that my own sister mocked, sending me the clipping of a newspaper cartoon once I'd arrived in Colorado that said, "So you want to find yourself? Save some time and money and just look in the mirror, stupid!" I hung it on the refrigerator.

While I didn't appreciate the sentiment, it's true that the choice was a shot in the dark and had a kind of romance to it. We were just cherubic little babies, so innocent and unworldly and unprepared for the commitment to each other and to this new adventure we had decided to take. We were sure to fuck it up. Who knew where we would end up both on this trip and in life? But that thought was what made it all so delicious to me.

We got as far as Colorado, where I got multiple jobs as a lift operator, snowboard instructor, and bagel maker during the mountain's mud season. I lasted exactly one year there, then left Colorado with a newfound love of Jameson's and the hard-won lesson of a first broken heart. Fuck it up we did. Though the clichés are impossible to ignore, I had also gained something harder to qualify—a more expansive worldview. By questioning the expectations I was used to and venturing into a new community with new points of view, I was introduced to new people, with new values, and new ways of thinking. Changing my *Surface World* opened my world (both literally and figuratively) and forever changed my *Inner World*.

Over the years since, I've had countless conversations with people who are desperate to know whether despite my success in the absence of a college degree, I regret it. Most are people still in awe over this comparatively clear decision by me. In those conversations I have discovered two things:

First, no. I don't regret my decision at all. Minus the broken hearts, I wouldn't change a damn thing.

Second, most people don't do what I did. Not the road trip. Lots of people do that, but not the pause before diving into a major life decision.

And that blows my head in because what happens when you *don't* pause is the stuff of my nightmares. I'm talking about the drudgery of a life lived by societal expectations—based on exterior influences and not one of free will. It is a life during which you wake up one day, look up from the conveyor belt, and think, "How did I get here?" David Byrne of the Talking Heads even made it into a song with lyrics reflecting that exact moment of waking up to the horror. When asked about it, he remarked, "We're largely unconscious. You know, we operate half awake or on autopilot and end up, whatever, with a house and family and job and everything else, and we haven't really stopped to ask ourselves, 'How did I get here?' "

We live in a fast-paced world in faster times than ever before and, if we allow it, our days become defined by stress, overwhelm, and a kind of autopilot. Escapism, automation, addiction, avoidance, and consumerism have drowned out the sometimes uncomfortable but incredibly necessary questions and feelings—the unease, the itchy brain, the sleepless nights, the creeping sense that you are off your track.

When I sat in the woods and got quiet, I was able to perfectly picture the life being laid out before me. I would go to college and receive a degree in journalism or communications. I would excel. I would marry my college sweetheart and take the first job that came my way in a neighboring city, and I would excel at that too. I would have a family and a dog and buy a house in a nice suburb, and that would be my life. Many people do that. It's a nice life. Hell, that's *almost* my life right now.

But in my pause, I was able to see that it wasn't the right path for me. I didn't want to rush into chasing down the staple trophies of life. For me, they felt like a trap.

When I bucked the expectations of my community, turned down that scholarship, and put my life through a cheese grater, it was a huge gift because it allowed all the good bits to remain after all the stuff I didn't want fell through the cracks. What I wanted was more freedom, more nature, more creativity, and more of the unknown. My inner voice wasn't mincing words. Fortunately, my upbringing fostered a belly of foolhardy guts to actually listen, when it mattered most, to what it was saying.

My path to success was not a traditional one, but it was authentic to me. I truly believe that's why it worked. My pause allowed me to see the silent but dangerous influence of expectations, and so I chose to drastically shift my trajectory, putting myself in the driver's seat.

To some, it may look as if I regularly and happily throw grenades into my life. What others see as land mines, I see as opportunities. A pause provides you with what we are all seeking: the ability to live your life deliberately. When we recognize that the source of the influence affecting our choices might just be a larger societal or community-accepted standard at play, it's crucial to take that extra beat to weigh your values outside of the expectations. The decisions you make afterward are in service of your true self, not the one sold to you.

Sure, my path—and yours—might be all squiggly lines rather than a straight road, but I've loved every twist and turn that has made me who I am today, here writing my book. No white picket fence required.

Folklore   What's expected of you is what you must do.

Folktale   Because everyone else is doing it, I must too.

Folk Rebellion   I understand that the people and community around me are under influences of their own, and I can recognize and weigh that before making a choice to go with the flow of a social standard, assumption, or expectation.

*Raise Hell   Think about a time when you made a decision without pausing. Look at the actions taken and the things you desired at the time against the backdrop of your *Surface World*—your communities' systems, structures, and accepted standards. Ask yourself if your choices would've been the same had you taken a moment, by stepping to the side of the conveyor belt, to gain a full-picture view. Now think about the next big decision you have coming up. What's influencing your decision-making process? What happens if you remove those expectations from your thought process—do your desires become clearer? How might your path shift if you remove the expectations from the decision equation?

CUSTOMS

(or Our
Outer World:
the Influences
Out There, the
Intangibles of everything
&Exploring
Perspectives)

For a while, when I was in my twenties, I became addicted to the feeling of starting anew. Changing zip codes for me was like trying on new clothing or different hairstyles and, in a way, new identities. Whereas some people might have feared that others would see them as flaky, I took pride in never living in one place for longer than a year or two. A boyfriend even gifted me a red T-shirt that read WISHY-WASHY across the front in puffy white velour iron-on letters, which I took as a compliment. My lack of commitment had become my proud identity, and the veiled insult became my new favorite shirt. Each move meant slipping into a different job and with that came new skills, new ways to view the world, new people. Somehow, even as an addlebrained twentysomething, I understood that switching my location meant changing my fundamental circumstances and with that came the opportunity to grow.

As addictions go, over time, I began to crave more extreme change. Hopping around U.S. zip codes wasn't cutting it; the buzz and stakes were no longer high enough. An increased dose was required—I needed to switch continents. So in the early aughts, I decided I wanted to try Europe on for size.

During a late-night dial-up-fueled search of the primitive World Wide Web from our shared family desktop (typically reserved for AOL Instant Messenger and Napster downloads), I happened upon the home page of an art school that accepted students regardless of college credentials. The upcoming semester would be split between the heart of Tuscany and a Greek island in the Mediterranean. If I was lucky enough to be accepted, I would study art history, creative writing, and photography. The website looked legit enough. The fact that they had one at all showed promise. It was the infancy of the digital revolution, and review sites, online press, and social media weren't things yet. I had to simply trust my instincts. The director's photo looked like what I envisioned an art teacher abroad should look like and described him as "an American expat," which, once I learned what that meant, I knew I now wanted to be more than anything. But most significant was this photo's backdrop—the coastal splendor of the Mediterranean. Nothing could have been more alluring to a girl who'd barely ever dipped her toes in the ocean. I applied immediately.

A few weeks later, via an antiquated Yahoo mail account, I received notice that I'd been accepted and was the recipient of a small scholarship to help in getting there. Very little additional direction was given. I was to wire the tuition money and board a plane to Rome. Just arrive by this time, at this place, *See you there!* It was a leap of faith.

I discovered quickly that moving abroad required more preparation than my other relocations. Maybe that's why fewer people do it. When logistics and money are a barrier to change, the ability to shift our location—and with it our influences and perspectives—is available only to those with enough resources or enough determination. It wasn't because I had money (I didn't) or because it was easy for me (it wasn't). It was because I wanted it so badly. I wouldn't take no for an answer when the passport didn't arrive on time, when my bank account wouldn't transfer overseas, when I got food poisoning a few days prior to my flight. I've found the best things are the ones you arrive at, on the other side, past all the places where most people often stop and pull over.

My passport became my most prized possession, a document I hoped would prove my gamble right—that (to me!) it was more valuable and powerful than a piece of paper signifying four years spent in the same place.

I didn't sleep much when I arrived in Rome. A travel agency had booked me into cheap accommodations—though I realized (a little too late) that the place was cheap for a reason. The hotel, priced in the seedy range (I now realize), was a hotbed for pornography viewing, featuring passionate Italian lovemaking for guests who rented a room. Just like many things preinternet, a liaison was required to get the things you needed. For booking a hotel—a travel agent. For viewing pornography—a "middleman" was necessary too. I just happened to be staying at one. I wasn't a particularly prudish twentysomething, but I was in a foreign country for the first time in my life and outside my comfort zone in more ways than one. I cried myself to sleep thinking I'd made the biggest mistake of my life.

The following day I arrived safely, but homesick, at the villa. In my quest for comfort, I began craving McDonald's. The only thing more American than McDonald's is an American abroad wishing for it. I was living in a twenty-bedroom Tuscan villa with four Italian *nonni's* (or grandmothers) cooking from the restaurant-sized basement kitchen and crafting epic feasts made from the handpicked tomatoes and olive oil grown on the rolling hills outside my window—and I wanted Chicken fucking McNuggets. This should have been a more obvious sign that I was majorly co-opted by an omnipresent fast-food and media-heavy culture—McDonald's and its admen had gotten their hooks so deep into me that I equated their products as an answer for nostalgia and home. But I have to admit, it took me a little longer to realize the cult of the cultural and media influences I was under from my *Outer World*.

In my pursuit for a taste of home, I'd walk a few miles down the hill to catch a bus to town. Getting to that bus stop was a grueling slog. As a resident of Upstate New York, and later the City, I was accustomed to walking long distances, but I hated the long stretch with nothing to do. Nothing to look at. Nothing to read or see. There were no bright billboards, no ads flashing by atop taxi rooftops, no flyers pasted to every bus stop and telephone pole vying for my attention. Just a stone road running alongside fields of olive trees. My brain was bored, and I didn't like the sensation that came with it. Old men were toiling in their fields while monks chanted in the hills above. It was a once in a lifetime moment, but my overstimulated brain was craving something different—the quick and dirty McDonald's equivalent of a walk.

In the end, what started out at first as a quick trip en route to the

McDonald's became a full day of adventures in the open-air La Sala Market, the place where locals have sold their goods and food and mingled in the piazza since the Middle Ages. Friendly vendors approached with pride to hand me fresh eggplant, hunks of homemade bread, and glasses of limoncello, suggesting that I "Eat! Try! Drink!"

I never ended up at the McDonald's.

In my own good time, I began to adjust. I found I enjoyed bus rides, gazing out the windows and watching the landscape roll by while I eavesdropped on conversations in a language I didn't understand. As my brain quieted, demanding less distraction, I soon opted to forgo my Walkman, choosing instead to listen to the lulling voices while embedding the postcard-perfect views into my mental camera roll.

In this corner of the world, in-your-face media and advertising were nearly nonexistent. Without the constant bombardment of visual noise fighting for my attention, demanding I spend my time paying attention to this and that, I discovered my longtime media-flooded mind had room to breathe. I could begin to think for myself without all the distraction. And without the stimuli I had been accustomed to in my *Outer World* back home, other things began to fall away too. The cultural trends, isms like can-do ism, and the frenetic pace that had pervaded my life started to show themselves for what they were. I stopped wearing my watch and relied on the bingbongs from the duomo bells. This little rebellion was a shift in dynamics from how I'd perceived the pace of life. Instead of something to spend or waste, optimize or fritter away, I no longer saw time as a powerful force silently ruling my life. At least while in Tuscany, I could let go of this norm I now realized wasn't necessarily normal everywhere else. Here, when the light shifted across the tops of the outdoor café tables, first cluttered with cappuccinos and then wineglasses, I knew evening was descending. Because I was no longer in a hurry, I delighted in whatever I discovered around every patinaed corner. My attachment to a fast pace slowly started to wane, and I looked forward to three-hour lunches, walks to nowhere, and long rides with nothing to consume except the present moment. I even began to look forward to my stroll to the bus stop.

I took in the rest of Italy with a fervor. I drank the Kool-Aid of *dolce far niente, the sweetness of doing nothing,* a term that would later be popularized in the United States by the book *Eat, Pray, Love*. The days lolled. Here, being unbound, unconfined, off duty, and free from so many things I didn't know I could be free from—stimulation, busyness, media, noise, productivity, advertising, and consumption— brought a sort of tranquility I hadn't ever experienced. I was alone but not lonely.

One day, on the way home from the internet café (yes, these were real things) where I emailed my weekly update to my mom ("I am alive, and I haven't married an Italian"), I passed a café that couldn't keep its party indoors. The people, live music, and dancing poured onto the street out front. It was in the middle of a weekday afternoon. I'd never seen such a spectacle before. These were just the type of people I wanted to be—or be around. Why banish fun to weekends when you could so clearly have fun anytime life calls for it? I bet they didn't wear watches either. Maybe I could soak up some of their *gioia di vivere, joy of living,* just by being in their orbit.

Returning with accomplices in tow, I danced in the café with the low-slung ceilings and on the cobblestones of the street out front. I smoked cigarettes as we attempted to communicate in half-baked languages over crusty bread dripping with oil. I learned to say "two more, please!" in Italian, *altri due, per favore,* having been introduced to my first espresso mixed with alcohol, and impulsively rode on the back of Vespas to another gathering, in another piazza, which we were assured we could not miss. Though it was well after midnight, children and people of all ages were everywhere. They splashed in the fountain, an accordion player provided the soundtrack, and old men sang and played dice games while couples danced. We'd acquired a few handsome friends along the way and promised to stay in touch. In broken English and body language, they begged us to stay and watch the sunrise with them. I would later learn that this is a regular occurrence in both Italy and Greece, at sunset and sunrise. People take the time to delight in watching it, and then literally applaud the wonder of nature. While that sounded lovely, it also sounded like a bunch of trouble in a strange place with strangers and no cell phone. As we asserted our desire to go home, the chorus of drunken pleas continued to grow louder until tomatoes rained from the balcony above in protest.

To this day, I like to collect moments like this one—the magic of intangibles you can't buy, you can't plan, you can't replicate or manufacture. This stuff often remains in the margins of day-to-day life when, in fact, it *is* the stuff of life. It's relegated to the periphery, not purposefully but because our *Outer Worlds* have taught us to cram our pages with so much constant content that there's almost no room left for chance moments, wonder, and unexpected delights. To pull the "margin moments" into focus, we must make room for them on our pages.

As an American living in Europe, I could suddenly see it so clearly, the effect my only known *Outer World* had had on me, though I could not yet put it into words. All I knew was that I had a desire for less of what I had known and more of whatever the magic was that Europeans had. My pages were jam-packed with routines, to-do lists, things to buy, shows to watch, schedules, productivity, and goals. Theirs appeared to be filled with laughter, calm, presence, enjoyment, improvisation, and awe. What I envied was their love of *just living,* just for the sake of it. There was no purpose to each minute, each moment, each day. Their *Outer World* culture wasn't constantly telling them (or selling them), that there had to be one. The purpose *was the day* and whatever unfolded within it. Their lives seemed to be ad-libbed, without purpose. Well, purpose the way I knew it to be defined—purpose as an accomplishment, something to achieve. But they *were* quite purposeful. They had intent in their playing it by ear, just for the sake of it, way of living. *That* was the whole purpose.

It wasn't until decades later that a therapist gave me the words to describe this way of being. It was an off-the-cuff and silly complaint. I used to enjoy the morning ritual of the coffee I made every day, but it no longer felt calming or provided a sense of enjoyment since my (very nice) partner would either make it for me or ask, "Have you had your coffee yet?" while looking around for something on the counter to tidy. To him the coffee was for the sake of something else—my ability to get my day going. I told the therapist that it made me feel stressed. My partner wanted to mark it complete as if a task had been checked off his mental to-do list. Making it something to be accomplished, it turned my much-loved ritual, which set me about my day with the headspace I desired, into something fast and forced. Instead of something I relished doing, it was a McDonald's version of grab-and-go joe.

She stopped me dead, asking him to explain when he felt the best— at *the end of the day, when I've completed everything on my list; when all the jobs are done, then I can feel at peace, put my feet up, and relax.*

She directed the same to me. *When I am fully focused on whatever it is I am doing, no distractions, no rush, just making a coffee, or working on my writing, going for a walk, or playing with the kids.*

"That's it!" she exclaimed. "You are process. He is outcome."

The gears of my brain shifted within it—finally matching up. With this observation, it all made so much sense. Life, America, him, me. This was the elusive key for making sense of nonsensical things.

She was very quick to point out that both types bring gifts to the table—*there is no one right way.* My outcome-dominant partner can help make sure things get completed with his get-shit-done attitude, keeping us on track to achieve the result we desire. And as a process-dominant person, I can make sure we find fulfillment within those moments leading up to the outcome we desire, even being okay, if the outcome we worked for doesn't happen, because the journey was satisfying enough.

To me, the only thing that matters more than the outcome is how we got there. There's no sense in achieving our "dream house" if it means that to achieve it, we've spent our time in a cycle of outputs that leave us distant, stressed, or depleted. I'll trade that dream-house goal for the joy of day-to-day life with a more attainable one (or no mortgage at all!) that allows for connection, calm, and fulfillment.

Our influences would have us think otherwise.

Our *Outer Worlds* determine a lot of how we operate within the world. History, economics, climate, culture, advertising, media, structures, and systems float around us all day every day, informing how we think, feel, and behave. It is because of my time in another place with history, structure, and culture different from the American capitalist society I'd known, that I was able to discover the frayed nerves and weariness I saw in ours, were not necessarily the same in others. My shift from output to process happened back then when my *Outer World* expanded, but I was just realizing it across from our therapist.

Looking toward the end of a journey, a life, a day—it's only human nature for people to focus on the destination, the achievement, the output. My issue isn't with the attempts to reach for things or having a focus on results but rather with the ideas that our capitalist culture has led us to believe—that there is an endpoint of fulfillment on the other end of a purchase, to-do list, or goal. When fulfillment remains evasive, because it is outside of ourselves, outside of the moments, and for the sake of something else, we continually believe that we must try again, buy again, improve a bit more to make the rest of our lives feel manageable. Maybe then we can feel peace.

But what if there was no end?

Studies have shown that in a classroom praising effort over performance, students end up doing better both academically and psychologically. If the goal isn't an endpoint (a good grade, a college acceptance, a job title) but is rather the process (learning to learn, understanding you can always improve, instilling a sense of curiosity, critical thinking, discussion, and a love of reading), then maybe fulfillment replaces disappointment, skills replace anxiety, and our *Outer World* influences don't train a generation to think that once they complete that certain output, they will finally feel fulfilled. Out of the classroom, it translates to understanding that what we do to reach something is more important than the something itself. On a smaller scale, it might look like taking a vacation with the motivation to post the picture on social media, remaining so focused on all the things to achieve that output that the vacation itself is secondary, and therefore less fulfilling.

In looking back, I realized that this was the best explanation for how I had experienced living in America compared with how I perceived Italians lived. We are outcome focused. They are process focused. To them, the process of living was more important than the outcome of living.

And so I went all in, making more room for the "margin moments"— not to tweet them or post them or live stream them. At the time, I didn't even call anyone to tell them about it. I would live in the scrumptious moment so I could remember it, be inside it, for as long as possible.

Living with gusto was not something that only happened in Italy, I soon learned. Once in Greece, Jack, our school director, showed us how to squeeze every drop out of every single thing. He skipped down hills, double daring us to see who could jump the highest or get to the bottom the fastest. We took walks foraging for herbs and ingredients to cook with. He led us on daylong hikes where we talked about art, politics, and philosophy, stopping to eat our bagged lunches in the sunshine. And like heathens, we stripped to our skivvies and ran into the deep blue Aegean Sea to float like buoys. The education was incredible, but it was these more subtle experiential teachings—*Get lost. Find the best espresso. Stop for the sunset. Always drink the table wine. Enjoy the serenade, even if it's inconveniently four A.M.*—that really stuck and have proved invaluable to a life well lived.

Before boarding the plane for my flight home, I placed a collect call from the airport in Athens to my father, who was very happy to hear I was alive, even though he was amassing international charges by the minute. Having traveled a bit himself, my father warned me that I might experience some culture shock upon returning. "I'm heading home, not leaving for some new culture, Dad," I said in a condescending tone, pay phone in hand. He let me have my ignorant say—and then laughed as he wished me easy travels.

Forty-eight hours later I arrived back in America. Cheap flights are cheap for a reason, I learned, just like cheap hotels. After six connecting flights and sleeping on airport floors in five different countries, I dragged my tired, dirty, tan, broke-but-happy butt off the plane and into New York's JFK airport terminal. I was so excited to be home.

Straightaway, the off vibes were undeniable. I wasn't yet halfway into the concourse, when something—no, *everything*—felt wrong. It was too loud, too fast, too bright.

My vision was flooded with posters, glowing logos, flashing lights directing me to buy something, watch something, or be something. My ears filled with the chaotic sounds of people yelling to one another over Ricky Martin and Jennifer Lopez pop hits. From the gate to the cab, I was accosted by magazine headlines, and messaging (*How to Win at Life!* or *Tackle My To-Do List!*), ads, and so much J.Lo— Jennifer with a Coke, Bennifer on the cover of every tabloid, the new fragrance Glow by J.Lo, and, of course, posters singing the praises of their new film *Gigli*. Fittingly, J.Lo's "Let's Get Loud" was playing in the taxi as I crested over the Brooklyn-Queens Expressway with the first views of towering Manhattan in the distance—and it seemed to represent the allergic reaction I was having to the stimulation, speed, and selling of the country I had known my whole life.

My instinct was to crawl into a corner once home, stick my fingers in my ears, and wince against the pace, commotion, and attempts at influence everywhere. My biology, my *Inner World,* went full fight or flight upon returning to the overstimulated American way I'd lived within since birth. It wasn't until I was back up to my ears in it that I realized I'd been free from the influence of the culture I had grown up with, and the accepted norms of constant media, promotion, and capitalism within it.

It should be obvious that different cultures foster different types of normalcy around influence, but just because one way has been sold as "normal" doesn't make it the best way of living. When I changed continents, unintentionally escaping my ingrained *Outer World* cultural influences for long enough for a perspective check, I was able to see my home through a different lens upon returning. I gained a new worldview, complete with an American capitalism detox, and it made me aware, whether I liked it or not, of how influential the culture you live in can be—and how easy it is to accept the norm in your sphere even when it isn't serving you. How easy it is not even to *see* it, never mind observe how un-normal it is until you learn that this is not normal elsewhere.

Moving zip codes was my drug of choice. Being strung out on action items and measuring up—it seemed—wasn't so self-inflicted. Although the Italian countryside approach to media more suited my preferences, over time the norms of America began to infiltrate and influence me again. That is how powerful our cultural influences are. Now, though, I was aware of its presence. And once you can see where the influence is coming from, you can begin to take back control.

In times of *Outer World* overwhelm, pausing allows us to detach ourselves from the noise, creating the option to either engage with it willingly once we've recognized and identified it, or make space, create, and embrace moments instead. Like my *gioia di vivere*-inspired morning coffee.

Folklore   Normalcy is good because it is normal.

Folktale   This is how it is, and so this is how I must be.

Folk Rebellion   I am my own person and will learn my preferences through discovery and curiosity about life outside of my own, choosing not to fall prey to the accepted norms of the heavy media and advertising culture just because I exist within it. Rather, I will intentionally search for what's out there for me, making room for the margin moments instead of being influenced by my society's structures.

*Raise Hell   What are some strongly held cultural beliefs or ingrained behaviors that you have? Consider how they began and became a part of your coding. Once you can see that this might be something you've just never looked at because it's the way that everyone around you also thinks or behaves, seek out a story or experience of another way—not in the goal of adopting that other way, but just to become aware that there are other ways. Now think about the things you do or feel that aren't serving you. Could there be another way, even though this is the culturally accepted way?

BITTER END

(or Inner,
Surface, and Outer
Worlds
Colliding &
Accepting
Bad Endings Are
Just False
Starts)

Being a starving artist in New York City is pretty damned expensive.

There are only so many couches you can crash on and dollar slices you can eat before you begin to question your life choices, especially when the quarters and dimes plucked from between the cushions no longer even cover the subway fare. It took all of a week back in the mecca of artists to admit to myself that I really wasn't one.

The truth is, attempting to join the creative class in New York City cost me much more than some sofa change and overtaxed friendships. It cost me my ideals and dreams.

It was a time in my life when I was trying to figure out who I was. I hadn't yet been able to recognize the influences of both the peers and places I surrounded myself with, or my own conflicting personal values within the culture, society, and structures I was a part of. No wonder I couldn't identify my true bona fide self!

Initially, I hit the scene hard with my newfound "I studied art in Europe" panache. There was a photography reception at a gallery on the Bowery; a friend's band playing at the city's oldest grizzled rock venue, the Bitter End; a painter acquaintance's open studio banger at his Tribeca loft.

I thought that to become one of them, I needed to learn from them. I had to know, *How did they do it? How did they survive and thrive?*

Over drinks and loud music, I would press, asking question after question, mostly about affordability and the likelihood of big breaks.

In my research, I learned that the photographer was being sponsored by her mom and dad, whose connections, in addition to their funds, helped land her an internship at a luxury travel magazine. The photographer's *Surface World* was far different from my own. Next! My guitarist friend lived with his bandmates, and countless other people, in a classic Lower East Side tenement, a grotesque and sticky version of today's social media content houses. If you had enough determination and Lysol, or were extroverted enough to enjoy sharing three bedrooms with more than ten people, it cost minimal rent. My *Inner World* was built with grit and determination, just apparently not enough. Moving on, not in! The painter was the outlier, having no answers to my "silly" questions, which I realized belatedly he'd never had to ponder. Having no idea how different he was—being truly gifted, outgoing, and also a handsome white man in a traditional boys' club—this elusive magical unicorn was the type of human everyone wanted to be or be around, especially in this scene. Our shared *Outer World* was a culture and society filled with systems and structures to lift him up before me.

Well, fuck.

I could've held out to try to become like one of them, I suppose: a classic black-and-white print photographer toiling in her darkroom with exhibitions at a Chelsea gallery. Or a writer for *The New Yorker* who chain-smoked cigarettes while rubbing shoulders with the city's other culturally elite mouthpieces at Elaine's, deciding fates over dirty martinis. Basically, like every other young woman in the city those days under the influence of HBO, I wanted to be Carrie Bradshaw. But in reality, these were just caricatures of the type of person I wanted to be. It was less about the making, the art, the creation, the writing, the having something to say and more about developing an identity and finding my place. Or more accurately, honing an anti-identity.

As I saw it at the time, being an artist meant I wasn't becoming who *they* told me to be—a college-funneled, sterilized version of myself on the career assembly line to indoctrination, riding the rise and grind to the nine-to-five in my sad cubicle, wincing at office inside jokes, at a job where I'd ruin the world just a teensy bit in favor of a capitalistic system that didn't at all give a fuck about me—every day for the rest of my life. I thought I was rebuking the pull of influence already by fervently rejecting society's expectations of me. But as we're learning, influence lives everywhere—and just because I thought I was rebelling against it from one sphere did not mean I was impervious to the other influence spheres around me.

But no. I was going to be different. My identity was going to be anti that. I'd be someone who woke when I wanted, created what I wanted, lived how I wanted, made less in favor of more, and had something important to say.

But how? If my scene-hopping in the world of artists and creatives showed me anything, it was that I didn't have the funding or talent to succeed in that world. Looking back now, I also didn't have a clue or anything notable to say yet. And maybe, just maybe, there was a part of me that craved more ease than that lifestyle offered—more comfort, more (*gasp!*) stability. Were my values changing so drastically? Or had I simply not been listening to my true self all along, crowded out by the influence of my artistically driven peers?

So instead of mirroring the values of my *Surface World,* I accidentally became what those people I admired called a sellout. My *Inner World* died just thinking about it. I had just turned thirty. I still had the rest of my life to become a well-regarded artist of some sort, I reasoned. But thirty seemed too old to live like a nomadic twentysomething on other people's couches. Some might wonder whether that was the societal influence of "timelines" or the concept of "adulthood" from the *Outer World* creeping in to sway me—good question! You're paying attention!

The truth is that beyond some self-awareness about my actual prospects as an artist and intellectual, I ultimately caved because of a quaint-yet-grungy garden apartment in the East Village that my sister and I had viewed while looking for somewhere to live. In my defense, there was a cherry tree in the backyard. What could be more perfect? I could see it all: the future this neighborhood would give me and the lack of proximity to the basic cookie-cutter existence it would also help me escape. My neighbors would be other deviants, defiants, and nonconformists. My new community (*Surface World*) would make it all more likely. My people—a mixture of poets and punks—who were not confined by societal expectations like marriage, children, office jobs, schedules, and the bourgeois trappings of success would be close at hand. This is exactly *why* I moved back to New York City at thirty years old, I reasoned. While NYC is in America, it's a far cry in many ways from the *Outer World* influences of traditional American culture. I was unmarried and childless and recently brokenhearted—just when I'd begun to give in to the idea that I might soon be both a wife and a mom. Where else culturally would the norms of being married and having kids by thirty be considered unfashionable and limiting? Where would being penniless, relationshipless, childless, and career-less at this time in my life suggest that I'd escaped? I so desperately coveted that apartment and all it represented on the corner of Avenue A and East Tenth Street. Living there seemed worth the sacrifice of a few temporary days at the office, so I began applying to marketing jobs on Craigslist, bent on selling my soul for that symbolic square footage—instead of honing my artistic desires.

I could use my powers—writing and storytelling, photography and creating visual identities, communication and conviction, a keen eye on aesthetics and trends—for bad, I reasoned. Only for a little while. Just long enough to get the apartment and change my spheres, my influences, and by default my opportunities—but only for cool companies. Just for the businesses with open floor plans and no early start times and unlimited vacation. Just temporarily. Just this once.

There was only one problem. Up until that point, my career trajectory had been all over the place—snowboard instructor, bike shop employee, ballroom dancer, fundraiser, bartender, continuing education instructor, advertising sales rep—and my résumé reflected that.

When I took an honest look at my work history, I recognized that there were two common threads: Each job relied heavily on my personality and my looks. Not something I am exactly proud of but a fact nonetheless.

Both my parents are charismatic, intelligent, and good-looking. I don't say this smugly. It's a look into how our *Inner World* influences shape not only whom we become but also the types of opportunism that come our way. According to the *Outer World*, my influences meant I had a leg up.

My father—a charming salesman, artful and sometimes

underhanded—taught me how to get out of a traffic stop, scalp tickets at Yankee Stadium, get—*and keep*—a bartender's attention, and read faces while playing cards. My mother—a former model, dancer, and Pan Am flight attendant—is still as gorgeous as she is good-natured. She took interest in other people's stories as a way to keep the spotlight off herself (and her privacy) and lived by the adage "You catch more flies with honey." As their daughter, I was clever while being crafty and believable while being persuasive. The conclusion was that—just like them—people liked me, trusted me, and listened to me from the get-go.

Proof in point (as they say in the marketing world)—the way I repackaged and sold my questionable résumé with bullet points instead of red flags. I designed it to look nothing like a traditional corporate CV, tied my mishmash of jobs together so they appeared purposeful and with a singular goal in mind: to become a marketing maven. I was so self-assured in my reframing that the résumé even had the words *Hey, there!* at the top instead of the standard letterhead including name, address, and phone. My confidence was palpable.

What I didn't yet realize is that my impulse—and unabashed willingness—to recast my experience in this way was the true indicator of my skill set. I had applied to these particular companies because the work seemed semicool and like it might play to my abilities. As it turns out, there is no job better suited to that cocktail of unscrupulous audacity than being a marketer.

I was going to become an artist after all—a bullshit artist.

I was hired within the week.

That was the moment it all changed for me, though it was death by a thousand cuts. This is a common term used to describe the way a major negative change happens slowly in small unnoticeable increments, making the change less objectionable, more palatable. The creeping normality of my new life—alarm clocks, commutes, calendars, cubicles, emails, emails about emails, work clothes, the use of soul-crushing office jargon (*Let's circle back and then touch base!*)—took hold over many small, unnoticed moments. My *Outer World* influenced my *Surface World,* forever changing my *Inner World.*

The thing is that that's how it always happens. One minute you're a self-righteous, virtuous, adventure-seeking nonconformist, and the next you're wearing kitten heels in a fucking WeWork, a modern-day mall of cubicles repackaged as a "mission," where unsuspecting patsies are pumped full of free beer and spa water.

There is a tool used in politics that I would like to repurpose for our learning and growth as we continue to recognize the influences around us. Picture a horizontal line with an arrow on either side of it. In the middle is *status quo* and *popular.* Immediately to the left and right of that is *sensible.* Another hatch past that is *acceptable.* And the next is *radical.* Hatches on the farthest point on the left and right of the arrow—*unthinkable.*

Now, if you were to look at my life on this line and put my *desire to be an artist* on the left side and *becoming a corporate overlord* on the right (you could also put left/right as creativity/stability or freedom/ money or identity/apartment—basically anything works, the idea is just to lay it out visually), you can see how choices over a period of time gradually shift to a place of acceptance or normalcy, which at one time might've felt uncomfortable or impossible.

The concept is called the Overton window, named after its creator, Joseph Overton, a senior vice president of the Mackinac Center for Public Policy. It is meant to show the range of what is politically acceptable to the mainstream population. While it works for political operatives trying to determine their policy making or what they can get away with, it can also be used on both a small personal scale and a very large dangerous scale. It's the day-to-day tiny slippery slope as shown in the kids' tale *If You Give a Mouse a Cookie.* It's also the large-scale slide of societal acceptance when inundated with propaganda and fear mongering, as illustrated in the "First They Came" poetic prose written about living during the Holocaust.

I think we know now that the window through which we think, believe, and behave shifts based on our *Outer World* influences such as time, culture, needs, policies, politics, media, and more. But learning that the window moves to the left or right, expands or shrinks, based on our influences and the actions, opinions, and decisions we take under them can be pretty eye-opening and empowering.

If I were to draw a box in the shape of a window on the line of my life at that moment, taking a dirty bed in an artist's house was unacceptable. And so my window began to shift to the right. In that moment it was more *acceptable* to me to take a "cool" corporate job as a means to an end. Over time, things I found *unthinkable* originally—using clichéd business jargon and spending most of my time in endless days of meetings—moved into the *status quo* part of my window as it—and I—shifted further and further away.

They also call this *gradualism*—as in gradually you go from swimming in your underwear to wearing Tory Burch kitten heels. Fucking kill me.

In attempting to shift away from both the influences of my art friends (*Surface World*) as well as the expected pace of life set by society (*Outer World*), I was more open to shifting what was acceptable to me (*Inner World*). Over a million little cuts pushing me further down that direction of my arrow, I ended up following the expectation of something else far away from who I was. I had adopted a set of false values, and their influence would be long-lasting.

As Carrie would say, "And just like that" I became the person I never wanted to be.

Folklore If you snooze, you lose.

Folktale I must follow the expected pace of life.

Folk Rebellion I understand that there are no timelines, no deadlines, and no same-pace-for-everyone way to do life regardless of what my *Outer* and *Surface Worlds* would like my *Inner World* to believe.

Raise Hell When life feels as if you aren't meeting society's (parents, friends, you name it) made-up deadlines, stop and ask yourself what would happen if you made a decision just to meet somebody else's timeline. Would you gradually end up further to the right or left of your window than you wanted? Even the best-laid plans (even plans not to have plans) can fall apart and lead you off course or, by leading you off course, can become a part of a much bigger course that you can't see just yet. Take the time to assess if you are doing something because you want to or because you are trading one set of influences for another. And if you have slid to the right or left (that's okay because we can and should shift our minds), was it intentional or unknowingly gradual?

The truth doesn't always win friends, but it certainly influences people.

—*Ferngully: The Last Rainforest*

Part II

How "They" Win Friends & Influence People

Magicians operate by a professional code forbidding them from revealing the techniques behind the illusions they create with common people. I'd like to say that there's a similar unspoken bond or pact among people in the business of influence, but modern-day citizens are so plugged in to this way of life, there's no need to hide in plain sight. And yet, we remain largely unaware of the devious tricks happening right in front of us.

In Part I, we explored *where* influence comes from and the importance of identifying where it exists in our lives so that we can begin to address it. You learned that influence can be inadvertent—like where and when you were born, who raised you, or what your position is in the sibling lineup in your family. These are things that are unavoidable from the outset. Influence that *just is*. But there are other types of influence that are intentional, premeditated, and deliberate. It looks to purposely point you in a direction. While not all influence is negative, dishonest, or deceitful, in this section I am specifically focused on influence brandished by entities that want something from *you* that benefits *them*. These influencers can be anyone or anything—an individual, a government, a brand, a headline, an app, a structure, or an institution. From here forward, I use *they* as a catchall.

My hope is that the information I reveal in the next section is so potent that the sleight of hand used by anyone intentionally trying to influence you for their gain is finally in the spotlight. Then you—the "target," "user," "customer," "audience," "buyer," "purchaser," "client," "patron," "prospect," "follower," "student," or "member" can become aware of their tactics—the insidious magic happening in plain sight—and no longer fall for their tricks. You'll learn to see past the smoke and mirrors through our next step: understanding *how* these influences work.

In our promotional and brand-saturated society, we have a problem. Well, we have many, but the one I'm addressing now is how we've become so accustomed to living in a world where we are constantly being told or sold that we no longer think it's odd.

To understand how influence shows up and how it's utilized, we first need to understand *how come?*

So, please, if you'll bear with me, I would like to work through some vocabulary and definitions, to clarify our influences so we are all on the same page before we move into the next part.

Let's talk about the four Cs.

To understand how to untangle ourselves from influence, we need to understand the four Cs because much *intentional influence* is employed for the benefit of one C or another. The four Cs live in allegiance to one another.

The first C, *capitalism,* is an economic and political system in which a country's trade and industry are controlled by private individuals and businesses *for profit,* rather than by the state.

The United States is the embodiment of free-market capitalism. Minimal government intervention and protection of private property rights, both tenets at the core of capitalist societies, sound great in theory, but what these pillars have enabled is the creation of colossal companies with little to no government regulation. Corporations have, of course, learned how to manipulate both the right to privacy and lack of government in order to exploit the people—*us.* The essential feature of capitalism is the motive to make a profit. I am not against the freedom to make money, but what we are experiencing now is a society and government that's been under the influence of capitalism for so long it favors protecting it above all else, even when it's being exploitative. In the age of late-stage capitalism, our economic system has shifted far from its original intent of benefiting individuals and small businesses—to big-business. Capitalism hurts our society. It is the system.

The next C is *commercialism*. This is the practice of trying to make *as much profit* as possible without caring about how it affects others. It is the spirit of profit above all else, a hyperfocus on maximizing and expanding—think franchises, consumer square plazas and malls, creating new customer bases, and brands showing up everywhere. Commercialism is the general attitude and actions of people and corporations influenced by the desire to earn money above all other values. Putting soda machines and iPads in schools even though we now know both are detrimental to the physical and mental health of children is commercialism. "Season of giving" advertising and promoting the collection of cereal "boxtops for education" are examples of commercialism. When gun manufacturers expanded their marketing efforts in the mid-1990s to focus on teenage boys, creating an entirely new customer demographic outside of the typical purchaser—hunters, outdoorsmen, tradesmen—it was commercialization. Commercialism is both overt and covert and is a direct result of a society deeply immersed in capitalism.

The third C (we are almost there, I swear; stay with me), *consumerism,* refers to the tendency of people living under the influence of capitalism to engage in a lifestyle of materialism. It's the idea that a preoccupation with the consumption of buying goods, a preoccupation that is reflexive and ever increasing, is desirable and normal. I am not talking about the essentials of living—food, water, shelter—but the economic motives to consume that developed in the twentieth century. With the commercialization of almost everything, removing oneself from consumerism is next to impossible today. In economics, there is a theory that spending by individuals on consumer goods and services is the principal driver of success in a capitalist economy. Essentially, without our being good consumers, our capitalist system fails. Consumerism is the result of commercialism, where the consumer is providing *profit for the benefit of a commercial entity in an exploited capitalistic system.*

The fourth and last C, *commodification,* is the one I really hate. Commodification turns things into commodities to trade or exchange. It gives these "things" economic value. A commodity is often defined as a raw material or something agricultural that can be bought or sold like sugar, oil, or gold. Today these commodities can be physical goods or intangible things like services, information, or ideas, but they can also be people, nature, animals, and art. Commodification is for monetization. Money. Sometimes we unknowingly commodify ourselves, as we do with the use of the "free" internet and online communities. A general rule of thumb to keep in mind in the digital world: When something is free—we are the product. In this instance, our data and personal information are being traded for profit. And, of course, there are some things critics of commodification believe should not be treated as commodities—water, air, children, wombs, body parts, information, education, humans—but the consistent "grow at all costs" culture of our modern world calls for new objects to become sources of value to keep up with the capitalist economy. Commodification is becoming the actual product for people to consume, for entities to commercialize and benefit from *to create profit* to feed the system.

Do you see a trend yet?

It took me a while to realize that at the root of most actions, beliefs, behaviors, opportunities, feelings, and motivations is one of the four Cs. When influence is exerted intentionally with one of the Cs as its guiding force, it's time to relook at that influence we have been living under. The only way to do that thoroughly is to understand *who* they are, *how come they're doing it,* and then *how they are attempting to do it.*

Once we know *who* the magicians are and *what* they're hoping to gain by their tricks, as well as *how* they do them, then attempts to influence us are obvious. Our eyes are opened.

For me, gone is the blissful ignorance of observing branding without a glitch, of seeing certain companies' hollow virtue signaling. I no longer perceive businesses, brands, industries, products, or employers as relationships in which they show me their higher purpose and have my best interest at heart. It is my understanding of the Cs, who *they* are and *how come,* that allows me to see through the tricks and tactics employed, to recognize attempts to influence even on the smallest of scales.

In the following essays, we will explore the path an entity takes to intentionally influence a behavior change—whether that entity is a friendly acquaintance, a politician, a lifestyle brand, or big media. In my brand marketing days, I quickly learned that the best way to persuade is to provide a solution to a problem. And then I learned that the problem could be entirely made up! Unscrupulous businesses have been making up ailments and "pain points" to "treat" forever. Understanding a person (data/demographics) and how they feel (pain points/key insights) to provide a solution (product/brand) to their problems while exploiting their emotions through messaging and communications (marketing/media) to trigger a desired behavior (action) is the well-worn path most commonly used in the business of influence.

It looks like this: *Demographic* → *Key Insight* → *Solution* → *Communications* → *Action*

This was the path I learned and used to influence. And so to help you on your path to becoming uninfluenced, I have laid out the following essays as such to help us uncover together exactly how all this influencing and trickery works.

Once the great seeing begins, you might feel angry or uncertain of yourself and start thinking, "How did I allow this to happen?" Take yourself off the cross. As you will see, the tricksters influencing your life are masters. *They* can be anyone or anything: a friend or colleague, a brand or business, a community board or church group, someone you date, and yes, the writer of a book. To become immune to their deceptions, it's imperative to learn the science of behavior modification. This is why I am taking you on a short and colloquial master class on the tricks of the trade, showing you how you're being sized up and cashed in upon. Don't worry, there's no homework, but you'll begin to learn through my experiences about the science of influence and the art of persuasion while honing in on the use of data, personality tests, stages of change, methods of framing, communications, and marketing funnels to create behavior modifications for praise, profits, politics, perks, and power.

You will start to see the attempts these influencers take to become "solutions" to the things that ail us, the very problems that are often created by them. Instead of being protected by our governments, as a culture under capitalism, we begin to accept the solutions of for-profit businesses as being temporary fixes. We are consumed by a steady stream of new treatments for the things that pain us—our ennui, depression, anxiety, overwhelm, sadness, loneliness—ultimately making us more reliant on the actual root of the problem. You'll see how we so often are persuaded to mask our issues with their quick fixes instead of actually addressing their causes—toxic social, cultural, and political systems based on capitalism—which then become "norms" because of the unregulated capitalism with no counterbalance.

Kind of like a parent who makes up fake symptoms—or intentionally causes real ones—so their kid believes they are sick, keeping the child dependent on them in a forced codependency through false ailments with false solutions. The child thinks they need the parent; the child thinks the parent is helping them. The child thinks the parent is looking out for their best interest. It's called Munchausen syndrome by proxy.

In the end, maybe it's not us who need to be fixed. While we've had a lot of self-care and wellness solutions, I'm not so sure about you, but I don't think we are very well yet. Even with all the nap pods, text talk therapy, mindfulness apps, antidepressants, outdoor apparel, running sneakers, and green juice, we really can't have the freedom to be well in a sick society that keeps selling us solutions to the illnesses they fabricate. Or maybe we were never sick to begin with?

Maybe it's kind of like those unwell parents, making us the unwell needy kids, who think they mean well? Dare I say it—are we all suffering from Munchausen by capitalism?

Just a thought.

So here we go: It's time for the big show! I'll be unveiling the tricks of the trade by those nefarious influence puppeteers in the following pages. Ready? Step right up to (and behind) the curtain with me!

WHAT'S YOUR FAVORITE COLOR?

(op Demographics:
How You're
Being Profiled
& Making It Work for You,
Not Against You)

My first two "big girl" jobs—or gigs that came with training manuals and a boss over the age of eighteen—were as a ski lift operator and a ballroom dance teacher. Orientation for each involved personality tests and discussions about the nature of human interaction. At the time, I assumed this was standard operating for onboarding new employees, but as I learned later in my career, not everyone who worked studied the way people work. It just so happened that the nature of those early jobs, and my future more stereotypically corporate work as well, benefited greatly from analyzing or molding our clientele's impressionable minds. Back then, I was just looking for jobs outside of cubicles, but what I unexpectedly received was an early 101 on the art of persuasion. Sometimes you get what you need.

The training at Winter Park Resort was well-intentioned enough. Management's hope was to devise teams made up of a balanced cross section of personalities, a strategy I'd later see work well for corporate teams and office culture. Throughout my career, my best bosses were smart enough to hire and surround themselves with people with opposite skill sets. The least successful leaders had more narcissistic tendencies, packing their offices with mini-mes and yes-men, which made for toxic environments with pronounced strengths but also severe weaknesses. So, looking back, it was pretty forward thinking in 1997 for a Colorado ski resort to throw a bunch of young, powder-seeking skiers and snowboarders into a room in order to size them up. Sure, we all had lots to learn in terms of honing our identities and skill sets. But we could learn from and complement each other. After all, the core of who we were, our natural temperament, was already coded.

That first day at the ski resort, they handed us what I would later learn was called the True Colors test. We sat in the orientation room with pencils and paper, ticking boxes next to adjectives that resonated with us. Were we curious or orderly? Warm or inventive? Spontaneous or parental? Eventually, we added up the hatches and scored each column like a round of mini golf. Voila! Our personality types were revealed.

When we're eighteen years old, the world tells us that we should already know who we are and what we want, but in truth most of us have no clue—nor should we. Adulthood is just beginning! So I couldn't wait for this sheet of paper covered with check marks to answer the big question: Who am I?

I've been obsessed with personality tests ever since.

This test is a simplified and drilled down alternative to the popular Myers-Briggs Type Indicator. You know, the one that boils down to shorthand letters that people now list in their dating profiles? Instead, the True Colors test likened each person's dominant type to a color, though it recognized that everyone possessed qualities of each one. I flipped over the scorecard and read: "Your true color is blue!" Words like *idealistic, romantic, personal, warm, spiritual, peaceful* popped off the sheet. Phrases associated with blue people included:

"I need to feel unique and authentic."

"I look for meaning and significance in life."

"I value integrity and unity in relationships."

"I am a natural romantic, a poet, a nurturer."

"I have a strong desire to influence others so they lead more significant lives."

I don't want to say this was a self-fulfilling prophecy—or perhaps just dead-on—but once again for the people in the back: "I have a strong desire to influence others so they lead more significant lives."

Each of the four colors, I learned, had very distinct qualities. At a macro level, they looked like this:

Blues are authentic, idealistic, empathetic, sincere, creative.

Oranges are spontaneous, witty, generous, competitive, adventurous.

Greens are analytical, calm, questioning, cool, focused.

Golds are punctual, organized, loyal, dependable, conventional.

Whatever else was the case, I knew right away that I was a tried-and-true blue. And not just because almost all my checked boxes lined up with that color. What was more glaring was what I was not. Even today I wish I could be more like gold. And yet in my honest self-assessment, I hadn't marked a single box for qualities like prepared, sensible, and thorough. More than twenty years later, this is still true, no matter how hard I try to flex, learn, or adopt those gold traits.

Knowing all this inspired me to seek opportunities and experiences later on that aligned with who I was naturally (or thought I was anyway) or balance the things that didn't seem inborn in me (my supposed weaknesses) with complementary people who held those qualities I lacked. It allowed me to be easier on myself for struggling with what I now recognized were not my organic strengths and accept that while I could learn to stretch myself, there might be someone better suited for a particular task. Internalizing this profile of myself was both validating and pigeonholing.

After reviewing the True Colors results, I considered, for the first time, how the nature of different "types" affected professional and social interaction. Originally, the personality profiling system was created in 1978 by Don Lowry, a student of the creator of the Myers-Briggs, to categorize at-risk youth into four basic learning styles, identifying the strengths and challenges of these core personality types. The resort employees weren't the target population, but I was definitely on the precipice of floundering, and understanding myself in more concrete terms felt like a lifeline.

Myers-Briggs, CliftonStrengths, Enneagram, Rorschach Inkblot, Big Five, DISC, Four Tendencies, Sparketype—after that first experience, I tried them all. Profiling myself and my traits was so appealing and helpful when trying to understand myself better. But all the tests resulted in a variation of the same thing: great at intuition, compassion, ideas, communicating, and loving. Terrible at planning, obeying, sticking to tradition, follow-through, confrontation, completing tasks, and logic. It all felt true.

But it wasn't until my next job that I learned how to apply this system for purposes outside of myself. Yes, I could realize my true nature in that profile, and that gave me a clearer picture of myself, a better chance of capitalizing on my strengths and overcoming my weaknesses, but it was when I hit the dance floor that I became adept at profiling others. I discovered that just asking someone to list the four color categories from favorite to least could tell you a lot about them. To this day, I ask friends and co-workers to do this all the time and then size them up accordingly. The party trick still works—never wrong.

Blue, green, orange, and gold. That's my personal order.

True Colors had told me that I liked anything "unconventional" and "artistic." So when it was time to return home from Colorado, I sought a job that would allow me to express myself and keep a flexible schedule. Options with those prerequisites in Upstate New York for someone without a college degree were slim until one fateful conversation with my mom. She was regaling my siblings and me with tales of her glory days as a ballroom dancer when she admitted, "I was hired as a dance teacher because I was pretty and personable." Check. Check.

Observing my interest, my mother dispelled my fears about not having experience: "Never danced a day in my life. They taught me the rest." Unconventional? Artistic? No experience or college degree necessary? Check. Check. Check.

I got up from the table and dialed the operator for the phone number of the studio.

As it turned out, the owner of the studio was my mother's very first student. Mom's student had moved all the way up to franchisee, owner of an Arthur Murray Dance Studio. And within a week, he was my new boss.

Brian was integral to my development. He taught me how to dance, sure. But he also taught me how to behave in the world. Things like how to enter a room, how to dress professionally, how to speak in a way that demands respect, and how to shake hands, make eye contact, and generally carry yourself.

I now realize that the version of me that Brian was molding was all about impressions and perceptions. Up until that point, I'd never seriously thought about how people perceived me. Sure, there were times when, like everyone, I was curious about whether I was liked or not. But I wasn't looking at myself constructively enough to analyze how I was viewed in order to modify or make shifts to change how I was viewed. Understanding how I was seen was the first step. Brian was teaching me to consciously consider how I wanted to be perceived in the world, but he was also coaching me to see myself as an extension of his business, studio, and brand, and therefore himself. It was also about how *he* wanted me to be perceived in the world.

Strategically choosing to be perceived a certain way, the way *you want* or that serves you, is one of the first steps in influencing someone else. The theory is known as impression management, popularized in 1959 by sociologist Erving Goffman. It is the theory that one tries to alter one's perception according to one's goals. His book *The Presentation of Self in Everyday Life* makes the case that impression management is an essential part of social human interaction because using it is key to influencing how we are treated by others. Because we all want to be treated well, we sometimes unknowingly or knowingly modify ourselves to achieve that goal. The attempt to influence the perceptions of others may be conscious or subconscious, but it's always about how individuals present themselves in a way to get what they want or need. By controlling the impressions that other people form of them, the manipulation allows the person to more easily satisfy their goal.

Brian wanted me to be seen as trustworthy, confident, knowledgeable, aspirational, and friendly so that clients would trust and ultimately hire me. You see, I thought I was becoming a dancer when, in fact, I was becoming a salesperson, an empathizer, an active listener, and an effective communicator—just one in a leotard.

This form of self-presentation Goffman likened to a stage play. Out front, where there is an audience, the actors put on a play for all to see. But backstage is where the costumes come off, the performance stops, and that privacy allows the people who were previously acting to act as themselves. Goffman theorized that in society there is a front and a backstage. I wasn't so sure about society yet, but in Brian's ballroom, I found it to be true. Because without being perceived a certain way by potential clients, I could not be persuasive with them. And then I would have no one to whom I could teach dance. I needed to sell those lessons. This wasn't a movement thing. It was an influence thing.

Brian trained me in these concepts with a specific intention. In fact, my days teaching dance lessons were bookended with our own lessons. Together, we would read books like *How to Win Friends and Influence People* and take practice tests based on the text. We would watch Zig Ziglar videos for motivation to get to the "real no" so we could overcome initial rejections. We would rehearse scripts for greetings and closings, putting nervous new students at ease and persuading them to commit to a starter pack of lessons. We would even enact a mock sale "close," learning which seat was the most powerful to sit in, how to turn our bodies to signal strength or interest, and even where to put our hands.

I also learned that my quick version of the True Colors profiling test could be used in reverse. If you knew the four-color archetypes and their traits well, you could usually guess which color someone was within a few minutes of meeting them. When they spoke, was their language fluid and all over the place? Or was it clipped and cold? Did they ask many questions, or were they overly excitable and competitive? When they learned the first dance steps, were they precise, wanting to know exactly where to put each foot, or did they want to feel the music and let their body take the reins?

I created an archetype for each color in my brain. Golds were my nemeses (in my mind). Type A's. Stuffy lawyers. Perfectionists. Brown-nosing class pets. Wall Streeters. Cubicle warriors. Think a rigid over-achiever like Henry Ford or some impossibly put together go-getter like Martha Stewart, who wears a white dress to the playground with her kids without staining it. And if she did, she has the stain stick in her purse to fix it. A gold would never take a blue seriously if a blue was being a true blue. A gold would think my language was loose and unrefined. Uncomfortable being led by feelings, they'd think all my talk about intuition and emotions was silly. My overt kindness would make me seem gullible and unwise. (I would later realize that instead of trying to emulate golds, I needed to be a blue and have golds in my life to save me from myself.)

Whether or not these assessments were accurate to a T didn't matter. What mattered, if I wanted to persuade people to, for instance, commit to more dance classes, was their perception of me. In understanding the essence of my color, and assuming theirs, I was able to modify my behavior to make myself more likable, trustworthy, and respectable—to *them*. So when it came time to sell them something, I could appeal to their taste and influence their decisions. And, boy, did I get good at it. By the time a prospective student crossed our chandeliered threshold over to the couch we sat at pre-lesson to build rapport and dig for barriers to "Yes" (kids, money, time), I'd already clocked them. If they were an Orange, I'd change to livelier and faster music to tap into their sense of adventure. Personally, I would give the appearance of being more competitive and wittier than I normally was. With Greens, I'd be sure to bring up what happens to their body biologically when they dance, hoping to spark their curiosity and engage in questions about the science of movement. I'd speak analytically and without emotion. If they were Gold, I'd pull out the clipboard and let them check the boxes on their dance card of items learned, allowing them to see the boxes left unchecked, a task I knew they'd feel the need to complete. In communicating with them, I would keep it direct, sensible, and moral. Blues were easy. I was just myself. Because of my profiling of them, and my altering of myself to be better perceived by them, I sold packages to Golds who swore spending money on dance classes was frivolous until I helped them see the long-term value of investing in themselves. I convinced Oranges to sign up for competitions to keep them focused on this hobby over other spontaneous hobbies that could take attention (and funds) away from dance. In no time, I became the studio's top salesperson and then, by the age of nineteen, one of the top in the nation. It was the role I was born to play.

Profiling people in such a close-up way allowed me to hone my abilities. It was in reading people and modifying myself accordingly that I was able to become a skilled observer of human behavior. Learning this party trick at such an impressionable age, an age where I wanted to get what I wanted, was like being given the keys to the kingdom.

Later on, I would learn that this type of profiling could be harnessed on a grand scale, zoomed out, with just a few data points about each human.

In today's world, we trade privacy for social media platforms and quick shopping so that brands and big businesses can profile and segment us. Based on their categorization (the boxes they put us in), we are delivered different experiences with their products. Each journey is customized based on their assumptions about us, gleaned from the data we knowingly—or unknowingly—handed them.

Having "key insights" into who their customers are and what makes them tick—whether through my rough-and-ready color assignments or more in-depth data analytics—is central to the way these entities try to modify your behavior for their benefit. We'll dive deeper into key insights in the next essay, but for now, just know that in knowing more about *you* via profiling, a brand or business or government can more effectively position itself—and communicate—in a light that seems favorable to you. Before the sparkly and fun internet was born, the archetypes of people (aka users, customers, clients, consumers, etc.) were a little more nuanced. We weren't clicking "accept cookies" alerts in hopes of watching that new cat video, allowing full access to our data so that a business could know everything about our interests. Back then, companies had access to standard demographic data like age, relationship status, and income bracket, which they'd marry with guesswork to categorize their marks. A business might have had targets like "middle-aged housewives," "traveling businessmen," or "college-educated young adults." Now that we happily trade our intimate information for access to memes, our data profiles might look more like this:

> Kim is a stay-at-home mom. Her husband, Dan, leaves the house each day by eight A.M. to work at his high-paying job in accounting. Kim doesn't feel supported and is regularly stressed out from taking care of the three kids, ages two, four, and seven. Kim is not looking to expand her family. To feel better, she shops online for beauty and fitness products and takes virtual Pilates classes. She has a hankering to

travel more, but the large mortgage and Dan's busy job don't allow for many vacations. Though she needs a real-life break, Kim spends any free time she has on Instagram and Pinterest, drooling over her friends' vacations and pinning the destinations she hopes to visit one day. Kim aligns with brands that focus on family, faith, and fun, like Disney, Jeep, and *People* magazine, but due to her declining mental health, she has a new interest in wellness. Though Dan is the breadwinner, Kim is the financial decision maker.

Why is this important? Because the more information these entities capture, the more it helps them shape their products and buying experience as a whole to better suck you in—not only so you buy more, but also so you feel lost without whatever they're hawking. The reality is that businesses on the forefront of profiling their customers, and leveraging those insights, outperform their competitors by 85 percent. This stat is conveniently provided by Gallup, a global analytics and advice company. The business of data collection is huge and growing exponentially. How huge? Facebook is said to collect more than fifty-two thousand data points (traits) on every user. This includes predictive analytics, which is knowing what users are going to do before they do. More startling, it's expected that the average thirteen-year-olds today have seventy-two million data points collected on them unknowingly. These metrics are then shared with companies seeking to target them. Gallup, a leader in the space for more than eighty years, is one such company making those connections. Gallup also happens to be the creator of the CliftonStrengths assessment, a personality test that twenty-seven million people have taken and 90 percent of Fortune 500 companies utilize.

You might be thinking, *Who cares? I like having a news and ad stream that caters to me personally because I've shared my information!* Sure, okay. I get that. I like Spotify's "Discover Weekly: Your mixtape of fresh music. Enjoy new music and deep cuts picked for *you!*" It's convenient and saves me the work of finding new music for myself, but what if a season of listening to Bon Iver after a sad breakup has Spotify suggesting an onslaught of Fleet Foxes and Taylor Swift?

Nothing against TSwift, but how does Spotify know I haven't moved on? Maybe now I'm in the mood for something more upbeat? Maybe I cry in the shower to Dr. Dre? Or there's a whole new genre of music I'd love if I wasn't being complacent and letting an algorithm choose my soundtrack for me? It can seem convenient to be profiled, but after a while, I get really tired of myself and want to expand my mind.

As much as learning about ourselves and our likes can be liberating, profiling can be pigeonholing and dangerous when someone else is trying to keep you in the box. Once you are put in that box, it's hard to convince these companies to let you out. I don't want to be codified as the sad single Brooklyn hipster mom (back when I might have been perceived that way). I want to decide for myself if I am sad or not and if TSwift is the best medicine. I shouldn't be influenced into liking a certain artist, brand, or product based on a season of myself or who-ever is paying Spotify the most to put their songs in the rotation!

Kim might think it's "*So Weird!*" that she was privately thinking about getting on birth control and then found herself reading an ar-ticle online about a new health service for women that can mail pack-ets of birth control pills without the hassle. It's not weird. It's predictive as a result of being profiled. Maybe it was the fact that three kids is the average for the zip code she lives in. Maybe she's been surfing a satiri-cal website, *Scary Mommy,* which has a focus on the not-so-great parts of being a parent. Maybe it was the bulk order of tampons from Target or the period-tracker app on which she keeps noting unpro-tected sex well away from the time of ovulation. We will likely never know which behavior made Kim a target for the pharmaceutical com-pany who knew enough to get their article in front of her eyes through the ruse of online "branded" editorial content.

Sure, I might be sizing Kim up. Old habits die hard.

Anyway, aren't we all being sized up or sizing up others all the time?

On a microlevel, we're being assessed daily by peers and neighbors, at home by the people we know, and in the world by the strangers we don't. On a more macroscale, we're being sized up based on how we search, read, watch, listen, what we buy, where we live, and our level of education. Assumptions are being made about our beliefs, values, per-ceptions, interests, attitudes, and behaviors, which can sometimes ac-tually frame and manipulate what our values, perceptions, and attitudes become. Just like the True Colors test, is it part correct and part self-fulfilling prophecy?

Sizing people up by their color code—or by data points—is like judging them by a single 280-character tweet. Yes, I can read people and sometimes it's fun, but it's taken me a long time to see that their color order is not the whole story. If we're all being generalized into colors or data points, what does that suggest about how large swaths of us are being manipulated at once and the way we, as a group, then influence one another to fit those stereotypes? These traits, assumptions, demographic data bits and pieces that aren't the whole of us are being used against us for nefarious purposes. Yes, people are all special, unique, and one-of-a-kind snowflakes . . . kind of. But instead of being treated that way, we are all effectively being reduced to easy-to-label categories. And if you're easy to label, you're easy to influence.

When I worked at the dance studio, Brian had a habit of calling me by the wrong name, so eventually, he nicknamed me "Trixie." Now, looking back on the career that in some ways he kick-started, it was as if my path of trickery was divined.

Folklore   We are all unique and can't be easily read.

Folktale   I am an individual, thinking for myself, in this bespoke world.

Folk Rebellion   I should understand myself so I can better protect myself from those who attempt to understand me.

*Raise Hell   Play with the competing dynamics of personality test assessments as tools to understand yourself better and also understand how others try to box you in. Consider the things that you've tried to get better at. What would it feel like to accept those things as they are, rather than optimize them? Appreciating that we all can't be good at everything, what if you spent your time and energy on the things that came easier or felt more innate? What would you keep and what would you let go? Now consider who has told you to get better at something (life, work, relationships, body, attitude, parenting, etc.) and why they were telling you to do so. Was there a structure, a product, or an entity on the other end that would've benefited from your "self-improvement"? Now consider the language utilized. Was it communicated to you differently than it would've been to somebody else with different "data points"?

MAKE BELIEVE

*(or **Key Insights:**
How Emotions Are Used to
Persuade &
Escaping the Echo of Ego)*

I've had an alias online for years. Today people call them ghost, fake, or burner accounts. These accounts are created to be disposable or anonymous, allowing the person behind them to stalk, participate, or troll unnoticed on the internet. In other words, they're sketchy, but I simply started my actual Facebook page, the one through which I collect real friends and connections, with a false name. It's been that way ever since. To this day, any connected login addresses me by my false online name instead of my own. Nothing could be a clearer illustration of how much our information is trafficked.

It's awkward checking into an Airbnb and having my kid say, "Hey! That's not her real name!" to the host who has addressed me as such. But the benefits in the early aughts outweighed the cons: I wanted to participate in social media, but I also wanted to keep my work life separate from my real life. It was an important part of my impression management. I was "me" on my Facebook page, and I didn't want to have to filter for a work audience. I didn't want it to become another space where I became conscious of how I was self-presenting. It was my backstage during a brief beautiful moment in time when that was actually possible.

Our digital footprints laid the bread-crumb trail of key insights for advertisers to learn who we were and what we wanted. Today personal brands look very different from the way they did back when I logged on to Facebook for the first time. Posts are curated and filtered and require approvals. Self-presentation of identity is at the core of the sharing now, with more tools and trends to help individuals show themselves in a way that gets them the rewards they seek (popularity, need for approval, or impression management) while the platforms get the information they need—your data. The internet is a smorgasbord where worlds, past selves, made-up selves, and people who should never meet collide.

As Facebook changed over time, its privacy settings and algorithm motivations shifted (as did all other social media), and my page became more visible and recognizable despite my wishes. I begrudgingly accepted friend requests from former teachers and old babysitters, disowned family members, and, yes, even a colleague or two—but just the chill ones. Then one day my Facebook profile and newsfeed began to do what everyone else's does—it fed me most everyone I knew and their random connections too. I've done a few epic culls over the years, unfriending people who don't actually fit my definition of a friend, but the number still sits closer to the size of restaurant seating capacity than those who actually show up to sit beside me in my real life.

The tech tycoons figured out early on that we humans are generally an ego-driven, narcissistic bunch, and so they built their business models and platforms based on that insight. It's simple: The way the algorithms work is that they give us more of, well, ourselves, so we keep clicking, keep scrolling, keep sharing, and keep coming back.

As we discovered in the wake of the 2016 U.S. presidential election, when the outcome demanded that we take a closer look at what we were being fed online, liberal-leaning people received more and more liberal-leaning news bytes and became more and more vocal about their liberal-leaning beliefs, which had become, as a result, even more liberal. Being surrounded by this information made liberals even more confident in their positions and more outraged, propelling them to post more, comment more, and share more about the state of the world. The right-wingers did the same but in reverse. They received more and more right-wing-slanted media, presented as fact even when it was far from the truth, making them more and more vocal about their right-wing beliefs. Assisted by the algorithms, depending on which side a person originally leaned, their perception of the world—their Overton window—shifted to the far end.

Instead of having one large window that encompasses most of our society and slides collectively together in one way or the other, we are being pushed as far to the left and to the right as possible. It's why people feel such a vast separation from those with different beliefs now versus decades prior. Political leanings weren't so far apart. Neither were the ways in which we viewed the world. It's a huge societal gulf in a way we've never had before, simply because there weren't tools to push us here. I am beginning to think that with the introduction of these profit-off-outrage algorithms, our single Overton window from which we altogether view our world has been fractured—into two.

Subscribing to, following, and watching media that mirrors our existing personal political leanings is quite normal, but it becomes dangerous when the tech echo chamber further reinforces our biases and beliefs. In communications, this is a media model known as the reinforcement theory. It operates under the assumption that people feel uncomfortable when they are wrong or when a previously held belief is challenged. To remove themselves from that yucky feeling, they instead seek out information that supports their pre-existing attitudes and ideals. Reinforcement theory is everywhere, but very widely used in politics. The three main mechanisms are (1) *elective exposure,* in which one chooses to only consume, watch, read what reinforces them; (2) *selective perception,* when a message cannot be avoided and doesn't match their beliefs, the person skews their perception to match their belief; and (3) *selective retention,* in which a person recalls only items that are in agreement with their predispositions, like a favored candidate's political messaging rather than their opponent's. Utilizing the idea that a person's behavior will continue with a certain level of frequency based on pleasant or unpleasant results, online platforms—and some presidential candidates—discovered that the more triggering the content was, the more likely a right-winger or a left-winger was to share it and espouse support for their side, thus triggering people on the other end of the (now more extreme) spectrum to share and espouse their beliefs, and so on from there. So begins the never-ending circle jerk of bullshit.

This is word-of-mouth marketing and commodification-of-people mixed into one. It's a big hairy sandwich of dangerous persuasion, behavior modification, and endless profit. The totality of shit, the end result, of these echo chambers remains to be seen.

Media conglomerates, politicians, and tech platforms were making bank at our expense, benefiting from our outrage and polarization. Ratcheting up the sensationalism and rage was great for business. But it was bad for us, the citizens of this country—our society, our relationships, our mental health.

In real time, I watched people in my life shifting their opinions and general behavior online to something much more extreme that didn't match who I knew them to be in real life. How many times did we watch in horror as someone who we thought of as generally level-headed began to share beliefs, clips, and theories that seemed objectively nonsensical? How many people did we see (perhaps) irreparably changed, to the point where their radicalization began to manifest offline and in person too? Even with my eyes more open than most, because of my interests and career background, I, too, was upside down and backward.

We all were being manipulated.

Regardless of how you feel about the candidates, the morality, or the outcome, no one can deny that the campaign Donald Trump and his people ran in 2016 was incredibly effective. It was one of the most persuasive initiatives I've ever seen—a veritable master class in influence.

Because of my experience in influence, persuasion, communications, and tech, I started to see the cracks in my reality in the weeks leading up to November eighth, and as a result, I wasn't as shocked as most people were that Hillary Clinton lost, and he won.

But I was heartbroken.

I've understood for most of my adult life that being the best doesn't necessarily mean the most successful and vice versa. An early immersion in sales will do that for you! If talent, effort, and intelligence were all it took to be powerful and successful, then we would have a government filled with talented, intelligent, committed people who earned their right to be there as opposed to those with the wealth, pedigree, connections, and/or social class to elbow their way in. Alas, this contemporary world is not a meritocracy.

This truth applies in almost every arena: We know that a product that's most affordable and reliable can come in second to one with better branding. But, also, a doctor who is the best in her field might go unnoticed because she doesn't hire a PR firm to help her make it onto one of the "best doctors" lists. A song might become number one not because it's incredibly good, but because it's engineered by audio marketing experts who expose us to it everywhere—in movies, commercials, and from the greased hands of DJs. The same goes for politicians. Being better qualified is not what gets people elected. It seems that self-aggrandizing people, mostly motivated by ego, fame, money, and power, are likely to put themselves and their families in the line of fire. For the most part, we won't get the best; we will get the best at self-promotion. And if by chance an honorable, selfless, intelligent person runs, say, in a Senate race, we won't necessarily vote for that candidate, choosing instead, time and time again, the one who is the best at selling—an idea, a feeling, or selling themselves. We will pick, purchase, and vote for the best at wielding influence. Almost always.

The actual best brands, products, ideas, and, yes, even leaders fail without the right tools for success. And the single most valuable tool for success—always, but more so than ever in our media-centric modern world—is the art of persuasion.

The most powerful and pervasive types of persuasion are inconspicuous. They are structures silently built into the foundations of government, entertainment, technology, and design that persuade, dictating our daily lives. Some are good, and some are bad. The most observable types of persuasion, though, happen through communication—speaking, writing, body language, and visuals like art and the printed word. Living in a community that has prioritized roads and driving, as an example, influences everything from how we interact with neighbors and our sense of community, to our physical and mental health, to the type of businesses that thrive or die within it. Unless consciously thinking about these silent influences, they remain in the background as "just a part of life." One city might try to passively persuade its inhabitants to become fit by creating influential structures within it like more access to walkways and bike lanes or communal gathering places and car-free shopping areas. The governing bodies of another place might take a more in-your-face approach to persuasion. Take, for example, the "Don't mess with Texas" slogan that even as a non-Texan you've most likely heard. It is meant to invoke a sense of Texan pride, of being tougher and stronger in a fight. Leaning in to Texan bravado, this messaging became a battle cry when it was slapped on bumper stickers and billboards across the Lone Star State. The reason? Litter. Based on key insights of Texans and importantly positioned as an identity statement, it made people (specifically drivers) feel a sense of honor about their home. This persuasive phrase reduced roadside garbage by 72 percent.

Words are powerful—especially when combined with the right images and insights. Creating strong mottoes and imagery is a tactic often employed for elections and during wartime too. You can probably close your eyes and picture the Obama Hope poster, designed by street artist Shepard Fairey to signal a cool factor and a connection to youth culture. Or the U.S. Army's classic I Want You poster, featuring an accusatory Uncle Sam, or the We Can Do It poster, featuring a strong woman in hopes of boosting female worker morale during World War II. Russia, China, and George Orwell's *1984* aren't the only ones to use propaganda. It's everywhere, regularly fueled by governments and media. And often the assumption is that it promotes a lie. But *any* information, true or false, that is created to shape public opinion and influence behavior is propaganda. It just has to be deliberate in its intent to persuade.

Of course those words on a poster must make you feel something to have power. That's where the key insights come in. And words on a poster that make you feel something are useless unless they catch your attention. Words on a poster that make you feel something and catch your attention are only impactful if they engage (or enrage). And it's all moot unless the message is crystal clear and actionable.

The most "successful" persuasion modifies another person's behavior and gets them to take action in a certain direction. In today's world, that could mean garnering retweets, clicks, or likes. It could mean prompting voting, donating, subscribing, or buying—all based on how we, as humans, are wired to react.

One expression that has saved me a lot of wasted energy in the last few years is "you cannot rationalize with an irrational person." But what if I told you that humans are not wired to be rational, no matter how much information is at our fingertips? And that that's part of what politicians and other influencers capitalize on?! In the 1970s, psychologists Daniel Kahneman and Amos Tversky blew wide open the way we think about thinking in their book *Judgment Under Uncertainty*. They conducted numerous studies that found that the specific filter through which we each view the world prompts us to make decisions emotionally that defy clear logic. Together they came up with eleven "cognitive illusions" laying the groundwork for what is today known as behavior economics. The illusions (our cognitive biases) showed flaws in judgment that happen automatically and without thought, to make decisions easier. Amos famously said of their research, "It's frightening to think that you might not know something, but more frightening to think that, by and large, the world is run by people who have faith that they know exactly what is going on." It's a terrifying thought and one which might explain why an unqualified person might think they are fit to run for office. If we can't trust the intentions, credentials, and logic of those who run, we must begin to trust ourselves, and only by knowing ourselves and how we operate (and how the influencers capitalize on that), can we begin to do that.

If logic doesn't sway and persuade us—what does?

Feelings. Fucking feelings.

Once you accept that gnarly fact, you can begin to understand how you're intentionally being triggered.

The "best" at selling and persuading already know this. They don't care about facts because facts don't matter if there's an emotion strong enough to overpower rationale like data, science, strategies—in other words, reality. It's how a religious person can cast a vote for a candidate who behaves in a way not representative of their religion's moral doctrine. It's what allows people to believe the claims that the 2020 election was stolen—because it feels to them like it was—despite evidence to the contrary. Perhaps most incredibly, during the 2016 election campaign, Trump not only convinced people to overlook his inexperience, but he convinced people that experience was a bad thing. He simultaneously bragged about being part of the 1 percent while claiming that he—unlike his elitist rival, Hillary Clinton (who used language in debates and speeches that was too detailed, unmemorable, rational, and unemotional)—was an everyman (despite owning, in the tower he named for himself, a gold toilet). With the shifting Overton window(s?), things once unacceptable became palatable. And through his campaign of fear, he validated the *feelings* of people who were already fearful, announcing, "You're not going to be scared anymore!" and then putting words to the previously unsaid fears, ratcheting up the language around them, thus altering their realities about losing jobs to technology and immigrants, about the radical left grooming our children, about criminal refugees crossing our borders in droves, and about minorities taking over our country. And then the people with the other Overton window did their job—sharing and shouting extremely left-wing media (like a headline on a social media feed) proving him right, in essence, supporting his statements by default. When people *feel* fear, it makes them hold tightly to their previously held beliefs, ideals, and way of life—turning away from anything unfamiliar, even if it is in their best interest. It was the *feelings* he stoked that persuaded people to vote against their own interests and values. Simply by making them afraid and angry, they were persuaded.

As a master of persuasion, Trump said what was necessary to evoke actionable feelings. People close to him have said he didn't even believe half of what he was saying, but rather said what needed to be said to elicit the response he wanted. It worked. It lit a wildfire of emotion in people (anger, excitement, rage, vengefulness). Facts don't matter when we feel something deeply. It's how yoga pants costing $138 are chosen over an almost identical but less "aspirational" pair for less money.

The Trump campaign was also effective in getting people to hold their ground no matter the course. No one wants to appear inconsistent or admit to being wrong. A person who is wishy-washy or flaky or often changes their mind is viewed as being untrustworthy, insecure, and even irrational to others. So that is why we humans, instead, tend to commit to one perspective and stay with it for just that social currency. We double down. Getting an influencee to commit to something, especially publicly, is liquid gold leverage. It's the psychological principle of consistency that sets our intentions in stone. We hate to admit to being wrong. So we dig in our heels and stay committed to, say, a candidate we posted we were backing on Facebook, despite realities arising that would have likely changed our mind. The Teflon Don often doubled down. He never backed off his actions or statements, no matter how clearly he'd missed the mark, so why should his followers? We'd literally rather change our internal beliefs in order to save face about our initial outward commitment. It trumps all else.

Our echo (ego) chambers are everywhere. With no one deciding to fix this incredible problem, it's once again up to us. And while it seems daunting to overcome a tech behemoth and its engineers who are triggering our psychological makeup, it's not that difficult to get outside of ourselves. Get offline, return to the real world, have conversations face-to-face. Listen to random radio stations and subscribe to unexpected newspapers. If we only take in what is served up to us in the clickbait hellhole of the internet, we will be persuaded down a path to the far left and far right with no middle window. The middle window can still have different takes, opinions, and feelings about it—but we are all looking at the same thing, the same view—which allows for those uncomfortable but necessary conversations. How can we even argue about a view if we aren't even looking at the same one?

Knowing we are wired to prefer what is familiar, becoming comfortable with discomfort is a great first step to not falling prey to our feelings or the algorithms missions. Had we realized this very human instinct to favor what we are accustomed to, to avoid the unknown, we might've speculated that in the 2016 election many would prefer to vote for someone who said, "Grab 'em by the pussy" over someone who had one.

Ultimately, the outcome wasn't as shocking when we look at the ways in which we, and so many others, fell under the spell of persuasion. The reality is that every time we sign into our social media personas (and the data profiles filled with key insights weaponized against us) and are upset or invigorated, playing into the division from which the campaigns, media, and tech companies are profiting, we're sacrificing a part of our very real selves. Whether we're using our real names or not.

Folklore   The news is simply news.

Folktale   I am free to have any opinions about it that I want.

Folk Rebellion   I realize that there are motivations for news, media, technology companies, social platforms, and politicians to profit or persuade me based on my biological human instincts and the key insights that my digital data provides. In knowing that, I can begin to think for myself by broadening outside of my own egocentric echo chambers, seeking sources elsewhere and becoming comfortable with discomfort.

Raise Hell   Thinking of the science behind why we do things, and the ways masters of persuasion use that science to modify our behaviors, recollect something you did or changed that seemed out of the ordinary for you. Have you ever done something or changed your stance on something without really knowing what caused you to do it? Can you see anything that might make you look at things differently now? Were you succumbing to clickbait? Were you parroting fear? Did you change an opinion because someone or something played your personal data (and supposed pain points) against you? Was the language being used a famous quote or propaganda? The infringement on our privacy, and on our biology, is everywhere. Can you see it? What Overton window are you looking out? What can you do to help create a single window for society?

HEADS BOWED, EYES CLOSED

*(or **Solutions:** How a Brand's "Good Intentions" Are **Bullshit** & **Learning** to See Through It)*

In 2012, when the Nets came to Brooklyn, I had to accept that my beloved Knicks were being replaced, in my city and my own home. I'd learned to root for the team from my dad, a dedication that was only cemented after twenty years as an adult in New York City. But facts were facts: There was a fresh, new, aspirational team in town—and they were *everywhere*. My son Hays was only one at the time, but I could see the future as clear as day. The lovefest I'd envisioned between him and the Knicks never even had a chance.

In marketing, there is a theory called the rule of 7, which posits that once a prospective buyer hears or sees a given marketing message seven times, they are more likely to take action (aka buy the thing, the solution). It's less about the specific number seven and more about the repetition of the messaging that's key. This rule of repetition applies not only to businesses trying to sell something, but also to almost anyone or anything trying to modify or influence someone else's behavior. Potential customers/voters/users/clients/partners/disciples/students/followers/lovers eventually go from potential to solidified once they become accustomed to brands/candidates/apps/practitioners/schools/etc. Simply put, if you're familiar with something, you're more likely to trust it and, therefore, adopt it. And the only way to become familiar with something is to see it, hear it, or experience it more than once. The old adage "The squeaky wheel gets the grease"—that is, attention is paid to those who demand to be most noticed—is at the core of influence and why marketing companies work hard to make it feel like suddenly something is everywhere at once.

So I guess it was inevitable that we became a Brooklyn Nets family. The allure of their advertising (and proximity) overpowered the sway of my history—for seven simple but inevitable reasons: (1) We live close to the Barclays Center, where they play, and members of the team live down our street. (2) The logo started showing up intertwined with Hays's first athletic love—soccer—with collab jerseys and televised watch parties hosted by soccer stars. (3) Everyone around us rocks the logo on hats, hoodies, and bags. (4) The playground where Hays has always played is sponsored by the team! (5) Game schedules are advertised on digital screens at our neighborhood bus stops and atop taxis. (6) The team ranks high on his NBA 2K video game, suggesting that it's one of the best. (7) We are surrounded by nonstop chatter about the Nets' iconic players, their trades, their shoe and endorsement deals, and their personal drama from kids, TV commentators, neighbors, fellow parents, and local New York City news. It's a vibe— a purposeful, well-strategized, and well-executed one. The reason why the Knicks were being eclipsed wasn't because they were outdated. It was because they were outbranded.

The Nets' allure goes beyond basketball as a sport itself. Sure, there are stats to memorize and games to watch. But there is also the culture of the club, the lifestyle being sold. The best in the business of influence know that you don't sell a product or a service. You start with a need or a want to fill, or a manufactured problem to solve. Athletic pants are a product to be sold. But people don't get excited to buy pants for pants' sake. They buy based on their wants and needs. Maintaining your health is a need. Looking cool in the gym is a want. Pants that embody that cool factor and showcase the pant-wearer's healthy identity, for example, are something people can get excited about (as evidenced by the $4.2 trillion wellness industry). Those brands that do it best create a feeling, a sense of community, a subculture clique around their pants (or whatever product) that becomes so cultish that customers (or influencees) begin propagating the brand's lure unprompted. They post pictures of themselves wearing the pants on social media; they talk to friends about how much they want those pants. This is called word-of-mouth marketing.

There are literally billion-dollar campaigns dedicated to creating "natural conversations" around products and brands. Nothing is more valuable than a layperson saying, "I loved that movie!" or "I'm obsessed with this new perfume!" When advertisers realized the power of word of mouth, they began creating stealth-targeted strategies to make those conversations feel organic in real life—guerrilla advertising, flash mobs, brand ambassadors, pop-ups. Social media has spurred the proliferation of this phenomenon online. It's how influencers and lifestyle bloggers came to be. Of course the truth is right there in the name—*influencers*.

At that point, it's not a product, a need, or a want that is being sold. It's a way of life, an image, a way of being seen. Buying into that way of life is how you end up spending hundreds of dollars on leggings (pants) in hopes of feeling like part of a cool aspirational wellness community, only to throw on that athleisure wear (pants) to sit on the couch or run errands. No offense to pants. Pants are just fine. But it's what you're unconsciously convinced the pants represent that can be a problem. Maybe it's a rung on your way to a healthier you. Or maybe it's a pair of pants that subtly reminds you that you as you are—is not enough. You can always be healthier. Maybe it's a ticket into the cool crowd. Or maybe it's a reminder that no matter what you wear on the outside, you still have to deal with how you feel on the inside, and why. Why do you want pants that might make you look cooler? Why do you care about being cool at all? (Psst, it's all about belonging. We will tackle that later, though.)

The Nets have a cool factor borrowed from Jay-Z because he was one of the original owners. They have street cred thanks to Brooklyn's unfussy raw reputation, but they also capitalize on the countercool factor: the borough's trend-setting street culture of hip-hop mixed with artisanal cocktails and famous hotdogs. It's not all fancy and cosmopolitan. The vibe feels real and fresh compared with other New York teams with longer, stodgier histories. The Brooklyn Nets are compelling because they're still being invented. The need the Nets fulfilled was a new take on old New York.

Supposedly, the team's logo was inspired half by craft pickle jars and half by parental advisory stickers—an ode to Brooklyn's rawness and constant transformation—hipster meets gangsta rap. It's rife with rebellion and flavor. It's ALL BLACK ALL DAY. Slogans like WE GO HARD! and BROOKLYN GRIT sell this image too, along with the riffing and re-invention of old-school titles like BROOKLYN STATE OF MIND. This team is not your parents' team. (That's quite literally the case for Hays.) Every color, catchphrase, and font was a strategic choice to make sure that difference was glaring. What could be more desirable than that autonomous swagger to a Brooklyn-bred kid?

Today Hays wears all the gear, obsessively watches the games on TV, and googles highlight reels online afterward. He begs for the streaming package as his birthday gift (never happening), takes clinics hosted by the team, and has the neon Nets banner hanging in his room, plus a framed jersey. His first heartbreak was the day his favor-ite player, Kevin Durant, was traded to another team. He fantasizes about playing for them one day, and he has already started buying his baby brother the same merchandise. Hays is now doing the Nets' bid-ding. In our home, it's nothin' but Nets.

None of this is a secret. It's happening in plain sight. In fact, when the Nets first launched, the CEO, Brett Yormark, a former NASCAR executive, was quoted as saying, "The goal is to become a lifestyle brand. We want to transcend sports. We want people to consume us as often as possible." This was said without irony. Without shame. With-out deceit. These businesses have no reason to hide their motives be-cause we are so accustomed to this blend of identity and product that the younger generations (in particular) don't see a divide between what they consume and how they define themselves. In their lifetime, one has never existed.

These companies have successfully branded themselves as cause driven or edgy in a way that people embrace without question. As much as companies say they want to "inspire," "connect," or "tran-scend" in their mission statements, more often than not their intent is to influence. We just have to see it. *Everything* a brand, business, or for-profit organization does is designed to get you to buy its stuff and buy into its brand.

Every. Single. Thing.

In for-profit businesses, they set up or partner with nonprofits, foundations, and charities to make you *feel* like your donations helped solve a problem or made the world a better place. They put these "good intentions" on their tags/soda bottles/websites/shopping bags/ (insert anything here). Water bottle companies donate a portion of their (net) sales to "saving the oceans" they're destroying. (Notice that they rarely say *what* portion. Also notice that they ignore the more logical option of stopping the production of their plastics altogether.) Food companies are lead sponsors of cancer foundations, while, in research, their mass-produced products are often linked to those same cancers. Beauty brands partner with mindfulness and talk therapy apps for teenage girls and women to help combat mental health issues like anxiety, depression, and eating disorders, while simultaneously feeding them a constant diet of beauty standards that can never be met—even with that two-hundred-dollar skincare system. And then there are the disingenuous "cause marketing" campaigns, designed to capitalize on whatever hot-button issue is currently in the zeitgeist, whether that be diversity, feminism, body positivity, equality, sustainability, and more. Brands all jump on the bandwagon. Or should I say, *brandwagon*?

Do-gooding in the form of reconstructing negative social norms, true allyship, making a difference through donations, promotion of issue solutions, trading bad materials for earth-friendlier materials— all make dents for sure. But the problems here are many.

Things get muddied when the good intentions are bound up in making money.

Translating our *feelings* (say, fears about the world and hopes for the future) into revenue streams keeps us, the purchasers, tied to the idea that things can be fixed through the buying of products. That "doing good" by spending money and acquiring things is the answer. And of course it keeps us tied to them—the brands, the profiting businesses, and, of course, capitalism.

Rainbows in shop windows are a common "season" in the consumer space now. Originally they were meant to signal allyship with the LBGTQ community during June's Pride Month. But too often now, rainbow products for sale inside have no relationship to the cause itself—a commemorative celebration to honor the 1969 Stonewall uprising (also called the Stonewall riots) in Manhattan and memorials to remember those in the community lost to hate crimes and HIV/AIDS. When the police raided the Stonewall Inn, a gay club in New York City's Greenwich Village on June 28, 1969, it ignited six days of protests by patrons, employees, and neighbors and is regarded as the catalyst for the gay rights movement that followed around the world.

Today the rainbow-washing of social media feeds and storefronts by brands and businesses has been co-opted to the point of a counter rebellion. In 2019, on the fiftieth anniversary of the riots, the Reclaim Pride Coalition held a separate march of forty-five thousand marchers to combat the "overflowing of corporate floats and sponsorship bribes." The members believe that what was once a cultural expression and legacy has been turned into an entertainment spectacle and branding opportunity that "had gone too far—too far from the spirit of the Stonewall Rebellion, and miles away from achieving societal equity for queer and trans people."

Buying a rainbow product because you are proud of yourself, proud of someone you know, or just want to support Pride are all totally acceptable reasons to purchase these products. Wearing our identities and causes as a badge of honor, statement, or signifier has been done forever and across everything. Sports teams, colleges, political beliefs, missions, religions, locations, bands, lifestyles have all created items to wear, hang, use, or collect to show buyers' allegiances. I have no issues with people spending their money on such things. But a seller who offers, say, a rainbow towel to a purchaser with the implication that the sale price contributes to the cause when there's no "good will" or charity attached is an issue I have. And that happens more than people realize. It's misleading the purchaser, tricking them to open their wallets, thinking they're making an impact when the only thing they're impacting is the bottom line of the business. Transparency is lacking when there's no evidence for where the dollars and difference-making are going—if anywhere. And even when a "portion of profits" is donated to an LGBTQ nonprofit, there is the issue of whether the business is really showing its stripes. Behind some of the biggest Pride-supporting corporations are even larger donations to politicians with oppositional views to such movements.

Even when it seems like a step in the right direction—when money mixes with morality—representation might just be another marketing ploy.

All this virtue signaling and social causes for capitalism, offering false or inauthentic solutions, can get confusing because of its deep hypocrisy. It's what enables a woke-focused, seemingly liberal brand to produce uniforms for Pride while its CEO makes moves in the opposite direction, or lets a male-dominated business pat itself on the back by sharing stories of its successful female leadership on International Women's Day, only to have the business's gender pay disparity published on Twitter hours later. It's how a company that creates and launches a new and supposedly sustainable product in honor of, say, Earth Day can still be a part of the bigger problem. The idea that consumers can create change for causes they care about through their dollars is based on myth. It puts the responsibility on the individual rather than the systems as a whole.

One fascinating move is to look at the mission statements of these companies, conceived to articulate in their own voice why the business or the brand exists. In theory, these should be composed of concise one-liners (or maybe two) and reflect the company's purpose. In practice, they invoke vague generalities and are often filled with touchy-feely words like *inspire* or *connect,* or vague goals like *potential.* To thrive in today's environment, a company's mission cannot be to sell pants and towels. Mission statements *must* be smoke and mirrors. The business must have a purpose to disguise the actual selling of said products and offer a fabricated solution instead.

A mission statement looks like this:

Nike: "To bring inspiration and innovation to every athlete in the world."

Coca-Cola: "To refresh the world in mind, body, and spirit, to inspire moments of optimism and happiness through our brands and actions, and to create value and make a difference."

Starbucks: "To inspire and nurture the human spirit—one person, one cup, and one neighborhood at a time."

Uber: "We ignite opportunity by setting the world in motion."

Facebook: "To give people the power to build community and bring the world closer together."

If Mission Statements Told the Truth:

Nike: "To sell sneakers, clothes, and equipment to generations over and over again."

Coca-Cola: "To repeatedly sell soda even if it causes heart disease and other sugar-related illnesses."

Starbucks: "To sell coffee in every neighborhood worldwide even if it shutters the local coffee shops."

Uber: "To profit from the service of transportation workers without having to pay for any actual cars or employees."

Facebook: "To collect everyone's data worldwide in order to sell targeted advertising, even to Russian bots to sway a presidential campaign."

The most successful mission statements leave a lasting impression of a brand's supposed purpose in the minds of its customers, encouraging them to choose this familiar name over another vendor time and time again.

Purpose is sold not only to consumers, but often to employees as well. It's touted so heavily that the culture of the workplace often revolves around it. It's painted on walls. Worn on T-shirts. Chanted at company retreats. When the company does a good job of this, sometimes employees develop such cultish devotion to that make-believe purpose (the front for selling pants or data or clicks) that they sacrifice important real things like actual social lives, building families, and general living. Fortunately, the pandemic brought a wave of wake-up calls for younger generations to the true motives of a business offering "free lunch" for their devotion as employees who had previously been sold the idea that they needed to have a purpose and that their purpose should be the purpose of the purposeful company. It turned out that life on the other side of the mission statement was much more fulfilling.

But we fall short of continuing that awareness of motivations when it comes to them and their products as solutions to all that ails us. We think nothing when a brand attempts to be our friend or even our savior. In fact, it's been made easy for them to slide into these roles as our trust in traditional institutions—organized religion, the government, our educational systems—continues to decline.

The Kearney Consumer Institute, a global consulting consumer think tank, released a quarterly study and briefing in 2021 denoting just this; aptly titled *Have a Little Faith in Me: the Truth About Consumer Trust,* that pinpointed "opportunity" in the continual polarization on all fronts. Kearney called this erosion in a person's ability to have faith in the institutions they typically turn to a "trust deficit." The institute goes on to encourage brands to take advantage of the empty hole institutions used to fill, by helping to resolve things they cannot on their own. They further note that brands can "fill the role of trusted advisor in specific niches since they're listening to consumers in a way other institutions aren't as much anymore."

We see this when a corporation offers to fly a woman to another state for a medical procedure that is no longer legal in the state she resides in, or when a brand hosts mental health practitioners at their wellness pop-up shop, or when a fitness brand class becomes our "third room"—the physical space where we connect with community in a meaningful way outside of our homes and work. When we start to turn to our brands to act as agents for everything that is failing us, we stop holding our actual institutions accountable and give these capitalistically motivated influencers a scary amount of influence over us.

Sure, we need things. *Things* like water bottles or pants or towels. These purchases will be made regardless. I am under no illusion that people will or should stop purchasing items for themselves. I love a good denim purchase, my weekly pickup of the "not the cheap one" bottle of wine, and my endless collection of books. Amazon cardboard has its own room in my house, and Revolve returns have made me a new friend with the shipping store clerk. And, of course, we are frequent purchasers of the latest and greatest of anything with a Nets logo.

I buy things. We all do. But what I do hope is that people start thinking about why, in their need for items, they are more drawn to some than others. Pinpointing whether this is an actual need, not an aspirational want, is one quick gut check. If it is an aspirational want—why? Has the company that got you to want that aspirational water bottle/pants/towel made you believe you might feel better in some way by owning it? How so? Has their marketing encouraged you to be better, live better, try harder? Or has the price been increased just enough to signal your socioeconomic status—or desired status—in actuality making it a visible totem of try-hards?

I buy expensive jeans because they hug me in all the right places. They make me feel comfortable. I purchase them from Revolve because their easy returns and orders make it possible for me to try many sizes and never leave my house. Books and wine make me feel like I am relaxing, escaping, enjoying leisure in a world that tries to speed me up. I no longer buy things that are marketed to me to try to be anything other than who I am. I'm off the improvement hamster wheel. And when I do try and "do good" like buy reusable water bottles, I also try to choose a brand that has clearly identified its do-gooding and has backed it up with transparency clearly stated on its website. And sometimes—I just don't give a fuck, recognizing that I just want the thing for no other reason than wanting it. I know when a brand swoops in to "help me," I'm most likely being failed in a bigger way by a bigger structure somewhere else. On a case-by-case basis, I can decide if that is a trade I am willing to make, and at least be aware of the bargain I am a part of. Sometimes my identity still gets tied up in a Yankees jersey or a band tee, purchased to show my love of my hometown and taste in music, but I am under no illusions that either will make me an all-around happier person. They are just shirts. It's what I try to teach my son, who's growing up in a time when, as one of the world's biggest advertising agencies, Y&R, declared, "Brands are the new religion. People turn to them for meaning." While sharing that consumer brands are now the thing that gives purpose to people's lives.

Dear God, I hope this world (and I) can give my kids more purpose in their lives than a pair of fucking branded pants can.

Alas, the Nets' formula worked. The rule of 7 lives on. Once he was old enough to make his own decisions, Hays did indeed choose the Nets over the Knicks. He will be a consumer for life. They found their audience and got him young. He and his peers represent a pipeline of profits for generations—marketing, branding, and advertising 101. All in favor of capitalism, commercialism, and his consumerism and commodification. And all this despite my constant lecturing.

My hope is that he understands that these are just pants, just businesses doing business, and his choices as he grows are still his, once he is aware of how the Nets, and every other brand, operate. If he wants to buy the branded sweats, so be it—so long as he is aware of their goal to ultimately just sell him something. His choice, then, would be one made with intention—weighed against his knowledge, values, and bank account. (Even though all the persistent brainwashing and ways they've infiltrated his life have made that choice pretty foolproof.)

Just think about how strong the Nets' tactics had to be to lure my son away from his family history and parental opinions—which are also, as we know, powerful influences. And yet the power of the Nets' solution, an all-consuming lifestyle machine offering an edgier, younger, more modern sports brand, overpowered my influence as his mother, an indicator of how entrenched we've become in this marketed world.

Hays went against the house, and then wore that brand as a Brooklyn kid who would do just that—with pride.

Folklore   Freedom of choice is everywhere.

Folktale   I get to freely choose what I like, need, and want.

Folk Rebellion   I recognize that my growing need, want, or like of something could be because my mind has been manipulated to feel such a thing, and I have the ability to weigh my choices and take action with integrity and intention now that I understand how this process works.

*Raise Hell   Assess the things you own. Throw open your closet and drawers and think about what that brand of milk or the label on the inside of your favorite jacket signifies. When you lace up your shoes or get in your car, are you experiencing the freedom or safety or vitality that your favorite modern brand implied you would? Notice the cost of something you own compared with other similar items. Has the value of what you spent delivered on its promise? Think of companies you've invited into your home, your healthcare, your child's school. Can you see any ulterior motives in their altruism?

MEGAPHONES & MARSHMALLOWS

(or Communications: How Mechanisms Are Used to Mold You & Regaining Control of Manipulative Mediums)

I don't know about you, but the first thing I do when I go to a person's house is examine their books. What they're reading and putting on display says a lot about them, whether that's who they are *or* how they want to be perceived. So when people come to my house and I see them looking at my prized collection, sometimes I ask them what it says about me too. Their answers aren't exactly the same, but there are consistent themes: that I'm interested in learning, curious, creative, intrigued by the brain and technology, not sold on religion, and invested in pleasure—travel, wine, and music. (I donate my guilty-pleasure paperback thrillers and mysteries about missing children and serial killers, but my "continue watching" stream of murder–y stuff couldn't be so easily hidden.) Curated book collection or not, it's an interesting litmus test in terms of understanding my, and their, worldview.

How we use media to shape ourselves, and shape our projections of ourselves, happens both on purpose and unknowingly. Sometimes it's subscribing to a highbrow magazine to strategically place on the coffee table while hiding the trashy gossip rag in the bath. And sometimes it's a new medium that's crept its way into our psyche without our realizing just how much it's molding us. Being selective of what media we consume is important because what it communicates will ultimately come to reflect us.

We *are* what we consume, in conscious and unconscious ways. For instance, if you watch reality TV, certain fashion trends might infiltrate your consciousness and perhaps lead you to follow certain people on Instagram. Maybe that contributes to your sense of what's an acceptable way to behave or where you want to vacation and, maybe, if you openly express that affinity to others, they might either bond happily with you or judge you negatively.

What others consume is something about which we make judgments—and while we don't judge a book by its cover, we're not wrong to assume that what a person absorbs does influence them. In addition to your bookshelves, show me the newspaper you subscribe to, your continue-watching stream, your podcast library, your browser search history, and I'll show you at least a piece of who you are.

The mediums we absorb—newspapers, magazines, television and radio shows, podcasts, books, movies, marketing from billboards to newsfeeds and, of course, everything we absorb on the internet—influence us. Our society prizes capitalism above all else, so its culture is fueled by consumption. And it's not just consumption of goods. It's also the consumption of content throughout these platforms.

I grew up as part of the MTV generation, a term coined and written about ad nauseam referring to those of us who were in their youth during the eighties to midnineties. As the term suggests, we were living under the influence of the first-ever music television channel back when it still played music videos and, boy, was it having an impact.

And now my MTV generation, Gen X youth, is being co-opted by the Gen Z "kids today." They're rocking nineties baggy pants and crop tops as indie sleaze. Grunge music—and its requisite flannels—is back. The kids are watching old sitcoms through their generation's eyes. And they love pointing out how inappropriate the jokes were, how bland the casts were, how they reinforced gender norms, rape culture, the patriarchy, and systemic racism across all mediums by the content that was created before rigorous diversity pushes and the social consciousness of today. Ironically, through television, magazines, books, radio, music, or advertising, the current generation gobbles up that outdated culture, with an arched eyebrow. They are explorers of nostalgia—a supposedly simpler time that, thankfully, demonstrates how far we've come. It makes sense to me why this generation—raised within an "everything happens so much" existence—envies the nineties. So if they want to borrow fashion trends and apathetic attitudes from my generation's less complicated era, I say, by all means—have at it.

The truth is, I love their malcontent.

For me, MTV was kind of equivalent to what social media is for the iGen kids of today. It was new, and we, the platform, and its content grew up together. In fact, as I write this, MTV is celebrating more than forty years of existence, entering middle age right alongside me. In my mind, it's declined in its old age exactly in the way our culture has. Since dropping the word *music* from the logo, it's instead become a home base for the documentation of spectacle and shamelessness. MTV, the pioneer of music television, was no longer about music. Music was no longer enough. Nothing was. Instead, they became the pioneer in reality television.

I turned away from *Road Rules* and *The Flavor of Love* reality shows that replaced the MTV videos I loved so much in part because the participants seemed to completely lack feelings of regret, shame, or embarrassment. I would cringe on their behalf. The Germans have a word for this sort of secondhand embarrassment: *fremdschämen*. Maybe I am a snob, or maybe this is my modern-day version of a *New Yorker* tote bag meant to signal my intended highbrow status, but I have opted out of the bad behavior fueling our industrial media complex that altered what's acceptable in our "windows."

I don't mean to blame *everything* on television, but its influence in our lives has made us lazy, selfish, materialistic, envious, indecent, shallow, violent, unimaginative, and numb. But it's a worthwhile form of media for us to do a deep dive into as we explore how the influence of media works both because of its pervasiveness in our society and its effectiveness as a medium of communication.

Canadian theorist Marshall McLuhan famously said that "the medium is the message." Television is a particularly effective method of influence because it is a mind-off activity. The goal of television is a sort of passivity. It doesn't take any effort to sit in front of it and watch it, which is why so many turn to it after a long hard day. By lulling the viewer, this delivery system is a particularly effective tool for coercion because it puts the watcher into a more suggestible state. Of course I'm less likely to blame the television itself for our shortcomings than I am to point fingers at the creators of the content or the corporate overlords who decide what stays and what goes—those who decide what we *like*.

In addition to shaping us, what we click, watch, read, and listen to also informs the money-hungry moguls of these mediums. Based on our activity, and the dollars they can make, they decide what lives and what dies, which in turn means the programming is based on what's popular instead of what's quality. As we've learned, something doesn't necessarily need to be the best to be the most successful. If it's packaged, messaged, and sold well, even the worst creation can be the most popular.

Algorithms push shorter and shorter content, making us progressively less able to focus on longer, more researched, in-depth, thoughtful, and nuanced stuff. Everything is a headline, perspectives only black-and-white, without dimension or gradation. Articles are shorter, videos are shorter, snippets of sound bites are digested for hot takes on three-second social media feeds or in 240-character tweets. Even our children are not safe from early influences that help to develop an inability to focus, embrace boredom, or even rewire their not yet fully developed prefrontal cortex. Animated shows oversaturate and become shorter, priming their young viewers to be good consumers of superficial and stimulating clips, videos, television programs, and movies in their future. Instead of watching a thirty-minute show on a static television with real puppets—the push to shorten "shows" on YouTube creates a slot machine effect for the child: What's next? What's next? What's next? Click. Click. Click. Skip. Skip. Skip. Ad. Ad. Ad. We seek faster and more outrageous content, making us feel frantic, which we *think* we want and need.

It's why we have the proliferation of shamelessness for fame, money, and attention, allowing *schadenfreude* to flourish. That is, instead of leaning in to natural feelings of embarrassment in the face of immorality, stupidity, or tactlessness, people consume the often self-imposed troubles, failures, and humiliations of others—with glee. And so our content gets more horrifying, more stupid, more unconscionable, and we become progressively desensitized to other people's mental and physical pain. It's a loss of empathy.

As a result, over time the calm and peaceful Bob Rosses of the world stop getting good TV time slots, or any slots at all. The next generation's David Attenborough might not even have a chance to make the next important natural history documentary if we continue to follow the model of the masses—shelved instead of shared. They're not high-octane enough to compete with spitting on people and crashing cars. This is our new unfiltered reality.

Without the interest, there are no eyeballs, and without the eyeballs, there is no reason to monetize and encourage more civilized programming. If our viral moments, or programming, could be based in kindness, compassion, and creativity so that we regularly consumed "smartened up" and higher-quality content, wouldn't we—personally and societally—become that, by benefiting from that influence alone? Turning our eyes away from the car crash of our humanity might restore it.

But because this is what we've been bred to like, the dumbed-down, stupefied, sensationalized, and satirized content isn't just for pure entertainment anymore. Now it's what propels our news outlets as well. In that industry, they have the saying "If it bleeds, it leads." Meaning viewership is higher when violence, death, or conflict is involved. To keep our attention on the screen, each story must be scarier, more spectacular than the next.

Slowly, we begin to live in an altered state, having been influenced by "news" that fans the flames of our unease and works to manipulate our perspective of the world. Take, for example, this misconstrued fact: The top three causes of death in the United States are heart disease, cancer, and respiratory issues. According to *Our World in Data*, the top three topics covered on the nightly news are terrorism, homicide, and suicide. What should you be more informed about?

You might be thinking: *If you don't like it, turn the channel. Unfollow that Instagram account. Read a different website.* Sure, yeah. I thought that as well. But the problem is that the message is no longer *just* the message when the *actual mediums*—not just their content—are influencing you as well. It is no longer just about what the media communicates, but the methods they use for communicating: the platforms, tools, and delivery systems themselves. There's no escaping the effects when the tool itself is rewiring our neuroplasticity, making our free will, well, less free.

Would I like to believe that most people, given the option, would choose knowledgeable nature documentaries over gaudy, narcissistic game shows? Of course. But that's just not the case. A few elements inform this reasoning. Time slots, which rely on availability and accessibility, are one example. It's similar to why a person will choose a McDonald's drive-thru instead of cooking a healthy meal at home— because it's easy, cheap, and en route.

When the majority of programming is anesthetizing—offering that escapism we now seek as a reprieve from this manic mentality the mediums have created—and the thoughtful, slower, deeper, and less easily consumable content is on in inconvenient non-prime-time time slots, on a network for which you need to pay extra, you end up watching what's been served up to you. And once something has been served up, well, that's how you get hooked. It keeps us sucking at the teat of their model. And as previously referenced, the more people who succumb to popcorn TV, the more the overlords' data (key insights) supports that type of programming (solution) from TV shows to clickbait (action). Simply put, if one stupid show was successful, the next show is likely to be far more brainless. And from there on, the algorithms exist to determine *how* to keep you watching. And, as we learned, it's more of ourselves. So one stupid clip you accidentally fall for will turn into fourteen stupid clips to binge watch, while suggestions for smarter content get buried. These "pop culture moments" are then turned into TikTok's dances, a product line, and the possibility of a new show in a new format as some person capitalizes on the notoriety, thus continuing the cycle. They need us to feed the beast.

In the early 1970s, a Stanford psychologist named Walter Mischel placed a marshmallow in front of a group of children, one at a time, and told them they could have a second marshmallow if they could wait fifteen minutes before eating the sweet in front of them. Ultimately, the study claimed to show that if they were able to delay gratification, it signaled a promise of success in the future. The children who waited had higher SAT scores, better education, lower BMIs, and more supposed signifiers of better life outcomes.

So are we all just kids sitting in front of a marshmallow tempted by instant gratification? If so, we're in trouble.

Thanks to the overlords and algorithms, we have no ability to self-regulate or choose what's good for us. We're so heavily influenced by the tools themselves and by the content of the media that surrounds us constantly, that we don't even know what we like or that we're only selecting from the options *they* are presenting to us. There is no such thing as delayed gratification in the world of media—especially in this time of on-demand everything. We have no restraint, no willpower. We're not supposed to.

Funny enough (or not funny at all), it turns out that even that damn marshmallow test was under the influence of the scientists' biases from the start. Researchers—NYU's Tyler Watts and UC Irvine's Greg Duncan and Haonan Quan—restaged the classic marshmallow test in 2018. This time, they cast a much larger and wider net, pulling in nine hundred children of more diverse races, ethnicities, socioeconomic backgrounds, parental education levels, etc. In contrast, the first test pulled its sample of children, fewer than ninety, from the same preschool on Stanford's campus (ahem—same zip code?), not taking into account all of these factors. (This is a perfect example of why it's not enough to read a quickly digestible headline instead of its full story. There is nuance to assessing information, and a distinction like ninety versus nine hundred kids is important.) The new, less biased test now suggests that the capacity to hold out for a second treat is mostly shaped by a child's background. A person's background, not their ability to abstain from a marshmallow, is what more likely determines long-term success or struggles in life. It's about the influence of our experience.

You with me? Lots of layers of influence to look at, I know!

So what does that say about our media consumption and the future of civilization? I might say, "Hold out for the better, smarter, richer content!" But what if that's easier to do for different people—with differing socioeconomic backgrounds and upbringings—than it is for others based on their opportunities to obtain or abstain? We aren't just talking about marshmallows as indicative of choice or restraint any longer. The marshmallow would be symbolic of our ability to make choices today only if the whole room were filled with marshmallows that had been connected to our brains and biology for decades to rewire us and our behaviors—to desire them, "jones" for them, *feel* we cannot live without them. And not just one. Each and every single one.

If you keep 'em connected, keep 'em dumb, keep 'em afraid, then *they* keep 'em coming, or rather, consuming. There's no impetus to encourage life outside the mediums. We've been conditioned to consume what's put in front of us. Marshmallows. Media. It's the same, but even so, that doesn't mean that there's nothing we can do. Here is something to consider.

Years ago, I read an essay in a book by George Saunders titled *The Braindead Megaphone,* which I think about often when it comes to the topic of media.

Imagine a party full of people from all walks of life. Into that party walks a man with a megaphone. He's not the smartest, the most well-spoken, or the most experienced person there.

But he's got that damn megaphone.

He starts talking about his love of spring. People turn to listen because it would be impolite to ignore him. Some agree with him, and some don't, but, because he is so damn loud, all the conversations in the room begin to revolve around what he's saying. The other people unwittingly adopt his cadence, his tone, and his ideas. If he says, "One hundred percent!" they start using that expression too. His rhetoric becomes the central theme of the party because it's unavoidable. He literally crowds out the other voices until the only one left is his own.

And just like that, Megaphone Guy's statements are now your thoughts.

The cesspool has changed drastically. We all have megaphones now, though they differ in size. There is power in this, of course. Like all mediums, our personal megaphones—social media, websites, blogs—can be used for good or for bad. Recognizing that might have us thinking differently about how we use them; we might also become more selective in choosing the megaphones we surround ourselves with. With everyone's thoughts following us around like a shitstorm of dings and pings, influencing our nervous systems, our attention spans, our emotions, our mental health, and our views on the world—realizing that these megaphones can come to reflect us can, and hopefully does, create monumental shifts in what we allow in.

Your mind is born with a blank slate, but your ideas, purchases, choices, and experiences in life are dictated by guys with megaphones. We live in a culture of commercialism, a world where news is entertainment, and marketing is based on the dumbification of the general population for the benefit of megacorporations that profit off you and your tuned-out, scared, easily distracted self. Recognizing this fact is more than half the battle.

If the Megaphone Guy is amplifying "unscripted" television focused on fame, money, and anti-intellectualism, and calling it reality TV, it's only a matter of time until we all tune in to that megaphone and begin processing the world through that lens. Our reality is now based on *Keeping Up with the Kardashians* (and their product placements), *The Apprentice* (yes, hosted by someone who became our president), *Survivor*, and salacious newer additions from *Love Is Blind* to *Naked and Afraid*.

When I look ahead, I'm less fearful about the apathy of the younger generation than I am about the coercion of the general population.

The digital mediums are taking the manipulative path with their addictive technologies and megaphones of lowbrow, often untrue, media, but we can push back against the quick-bite influence being pushed at us. One option—if the medium is all-around unhealthy for you—pick up a completely different one: a book, a (*gasp!*) paper magazine, or listen to the radio. Not always, and not forever, but removing yourself from the medium for a bit can offer a perspective on how it makes you feel. Another option is to remain on the medium but push back by striving for slow moments. For every reality show, a streaming service pushes at you (and maybe you like those, which is just fine if it's *your* choice!), challenge yourself to watch something new outside of your usual wheelhouse. For every six murder paperbacks you read, try an essay collection or weave in a nonfiction how-to book at the same time. For every TikTok marathon you get stuck in, go for a walk outside.

Maybe by doing so we can begin to turn away from the spectacles we've normalized. If we continue to live vicariously through our mediums and the content on them, because we are what we consume, we should choose to turn to the stuff that allows us to find delight in others' joy and happiness. By not watching the car crashes of our society and giving our attention to the stuff that breeds empathy instead—we can make a bigger difference than we realize. Just by choosing what we opt in or out of, we are telling mediums what we want, what we think is acceptable. Remember, the overlords are always watching.

Folklore You can hear, listen, and watch without being influenced.

Folktale Media and technology are just parts of everyday life.

Folk Rebellion I recognize that the messaging techniques that influencers use to communicate with and exploit me can take many forms, and that there are different delivery systems and different types of media and different types of quality and content. I also recognize that each one has a specific motivation. I have multiple choices in what I choose to consume for entertainment, news, information, or education. I can choose which medium and how often I use it. I can choose the types of information I allow into my brain and life, and dare to think about my media the same way I think about other things I consume, like food, alcohol, clothing, etc.

*Raise Hell Look at your media and try to picture what it looked like a decade ago. Was there "still watching?" on your screen? What about longer? Were there DVDs in the mail? Take note of how the mediums you use have changed over the years. Do books on your digital readers feel as relaxing as ones with pages? Does your need for the news keep you on your phone into the late-night hours when it used to be on your TV at five o'clock? Check in on your attention span. Has it felt itchy, shortened, sporadic? Could this relate to any change in the media messaging you absorb in your day to day? Think strategically about what you consume mentally. How much is mind off versus mind on?

WEIRDOS

(or Action:
How It's About
Yours, Not Theirs, &
Challenging
Ourselves
for a New Way of
Being)

There was a half-eaten french fry hanging from my mouth the moment I realized I had both saved—and saddled—my son.

It was during one of those lunches with fellow Brooklynites—"parent friends"—where over beer and burgers we try to act out some charade of who we were and where we belonged prechildren. Our current reality was spilled milk, broken crayons, tears at the arrival of a *no*-cheese cheeseburger *with* cheese. I don't know whether we continued to make these fucking miserable, hopeful attempts due to parental goldfish brains or a longing so deep for our old identities that we kept returning in search of a beer we *might* actually finish this time and a conversation we *might* actually be able to have. Our white flag of surrender, the tab, thankfully arrived, signaling the merciful end of our exhausting "lunch."

I was paying the check at the bar when my son noticed the television hanging above it.

"Look, Mom. There's the advertisement Dad loves."

He was five at the time.

Some might think this was an odd observation, and a sophisticated word, for a kindergartner. My friend Kelly definitely did. I knew he was being conspiratorial with me, ribbing Hays's dad in absentia for being a sucker because he knew it was something I would say.

Misunderstanding his tone, Kelly interjected proudly, "Well, Hays, did you know that I make advertisements for Facebook?"

Oh. No.

In our house, I had taught Hays that Facebook was bad. I'd told him that despite being everywhere, advertisements were nothing but tricks, lies, and bullshittery to get him to buy something he didn't need, do something not good for him, or believe in something not true.

Instead of grinning back, Hays made a face looking like he had smelled a shit sandwich. My fears were answered. Five-year-olds have no poker face. Without hiding any of his disdain, to my amusement— and Oh-my-God-could-the-earth-open-up-and-swallow-me-whole embarrassment—he snarled, "*Why would you ever do such a thing?*"

Poor Kelly. I covered my eyes.

As we covered, how we are raised and by whom is one of the greatest influences on each of us.

And this was my son.

Of course he would know about advertisements. I—a nonconforming skeptic, a questioning iconoclast, a subversive troublemaker, a repentant marketer, a real know-it-all with a big mouth to match—was his biggest influence. I made sure of it.

As Hays's mother, I'm the narrator of his childhood. He breathes in what I say, do, and believe automatically, like air. So to people who know me, it's no shock that at the time he referred to God as *She*, believes the internet is "the man," working is trading your time for dollars, and big houses and big cars are unnecessary time trades unless you have a big family. Today, he also believes, after many years of conditioning, that being weird is good. That feels like my biggest success.

One of the greatest drivers for us as humans is belonging. It's a primary reason why we enter relationships, join clubs, care about what the neighbors think, root for certain sports teams, and practice religion (or don't). It's in the clothes we wear, the cars we drive, the music we listen to, how we take our coffee, and how we raise our kids. The need to belong is about more than knowing others and being in community with them. It's about social acceptance. And that need is so powerful that gaining approval from the group can lead us to change our behaviors, values, and views to conform to a societally accepted way of thinking.

As a result, trying to avoid being perceived as an outsider, a misfit, a weirdo—defined negatively by most—is one of the greatest universal influences. At the core, all we want is to belong, and that means adhering to social norms, not bucking them. That's why every major brand now has a "community" on top of their product or service, whether it's a run club, an event gathering space, or workshop. There's a positive link between a sense of belonging and greater overall happiness and well-being. On the flip side, a lack of community increases mental health issues like loneliness, anxiety, depression, hopelessness, social anxiety, and suicidal thoughts. The iGeneration that grew up in isolated play indoors, and mostly online or on phones instead of meeting at the local rec center, have shown a drastic increase in all these issues. Belonging is that influential.

I hoped that if I could influence Hays to think that being weird was good, essentially repositioning the term, then I could potentially save him a world of hurt (and fruitless conformity) for thinking differently. Maybe he could feel as if he belonged without having to completely conform—about basically everything: style and clothing, notions about the "right" education and career, herd mentality when it comes to views in pop culture and politics. I knew my sway would sometimes lose out to other influences—from Nets marketing to objectively terrible country-rap hybrid songs—but maybe I could save him from some of those useless "shoulds" and moments of insecurity because I removed the stigma from being a freethinker!

In the case of Hays, my influence was somewhat inevitable. As his mama, I am both his authority figure and caretaker (who he thankfully likes very much). And those are two key components in any dynamic of influence—authority and likability.

Likability is the wonder drug above all others. Often the words *charismatic, charming,* and *socially savvy* are used to describe masters of influence, the foremost leaders, gurus, C-suite executives, and celebrities of our world. It's not necessarily that they deserve those positions, so much as that they have leveraged their likability—a trait that's often hard to define—to get where they are. Often it's less about the traits a person possesses and more about their ability to make others feel special, playing into the ego echo chamber. Whereas, earlier, we learned that those wielding influence played into *feelings* of fear, in this case, the influencers are playing into our need to *feel* special or to belong. *Feel* important. *Feel* interesting. *Feel* seen. Whether it's innate or cultivated, their higher-than-average emotional intelligence (or EQ) makes them particularly skilled at shining their light on others.

If you and I have met at any point since my late teens, when at the ski lodge and dance studio I studied the way the mind works and what intrinsically motivates humans, I probably know more about *you* than you do about *me*. This is a skill you'll often see in master influencers. See, ever since I discovered that most people's favorite topic is themselves, I've learned to ask questions and listen rather than take up too much air space. In my later years as a brand strategist, I was able to parlay this knowledge into a bona fide force field for myself and a treasure trove of assets for me to leverage. For example, when working with clients or co-workers, disarming them enough to open up for the duration of a three-day on-site conference, I kept them divulging, sharing, and being honest. And they shared just about *everything,* including who really controlled the budget (which happened to pay me), which middle manager would get the axe in the next quarter, and who said what about whom. Often the information was valuable, but more valuable was the leftover mystery of me. Like the comedown from a first date gone too far, I could only imagine that my targets were revisiting our time together and might be having second thoughts as they realized that they knew not very much about me, but I knew a lot about them. The "real them" began to show up over time not only in what they'd told me about themselves, but also through what they'd been willing to share (were they someone who gossiped, someone who had no lines between home and work life, or someone who let their hair down too much?). By making them comfortable, I was able to take on an authority role, albeit a friendly one. I was now a leader, a person in charge nonetheless, by tipping the scales just enough.

What started out as a learned skill that I developed in hopes of influencing others soon became a social crutch, leading me to smile, offer personalized compliments, and tailor my observations to a person's interests during every interaction. Eventually, I realized that even though I was no longer attempting to sway or persuade, I couldn't shut off my role as a vulnerability deflector. I'm so good at conforming to people's needs that sometimes it's challenging to show people the real me. It simply worked too well for me. (Thank you, therapist!) Quite a lot of the skills I learned back then are still coded in me and have fundamentally altered who I've become. As my sisters like to say about me, I'm "always on."

And it doesn't just stop at smiling and listening.

I confirm what I've heard by repeating it back and then validate any feelings people express to me by empathizing—full active listening. If I point out mistakes or different perspectives, I do it indirectly through stories about my own experiences. If I have something bad or negative to share or correct, I deliver it in the form of a compliment-shitsandwich—positive first, then the negative, then another positive. Voilà! I always bring the conversation back to the other person and I always try to leave people feeling as if they've come to a solution themselves versus me telling them what to do. I volley up quandaries and questions that lead like a bread-crumb trail to where *I want* them to land.

In a way, this book is built just like that.

The types who are best at enacting change read people, adapt to meet them on their level, and communicate in a way that will be well received. They are personality chameleons in how they morph in their likability to match their target just right. This ability to make others *feel* comfortable allows the influencer, writer, leader, mentor, TV evangelist, teacher, or car salesman to appear more authentic and, therefore, more trustworthy.

It's an intuitive tuning in to others.

Hays's kindergarten teacher once told me during a parent-teacher conference that she considers him the "antenna" of his class: the weathervane, the barometer. She checks in with him, to check on the rest, because he's so "tuned in to how others are feeling." Hays's function as the temperature gauge of his class might make me feel warm and fuzzy at first blush. What mother doesn't want to know that their kid is attuned to the needs of others? But holding the core power of likability—an intuitiveness when it comes to others—isn't something

Hays has realized he has yet. Whether it's a good or bad thing that he has this skill hasn't yet been determined. That won't happen until we learn *why* he's so good at it and, of course, *how* he chooses to use it. In one vein, maybe he just cares a lot, as a natural empath, or in another, maybe he just wants to belong. Often that desire is the first stop on the road to perception management. He may have a need to fill. The influence of wanting to be liked, to belong, is so deep it can spawn people pleasing, fawning, and conforming, but that means a focus on always being tuned to how *everyone else feels* instead of *how you feel yourself.*

The coding continues . . .

Now you might be thinking: *But what about those really successful people who seem like horrible human beings?* Good spot. In the absence of likability, there is authority. And authority can be faked, smudged, and exaggerated in a million different ways. It's what paves the way for the rash of cons being peddled, like a luxury music festival on a private island that turned *Lord of the Flies* or a German heiress who successfully fundraised for a landmarked six-story $135 million Park Avenue address for a "club," despite having no backers, no money, and no professional experience in art or clubs. It's what allows an Elizabeth Holmes to attract venture capital funds with a PowerPoint presentation that sizzles but has no viable product. All it took was a convincing authoritative vibe. She even intentionally lowered her voice to become more so. Every asshole on the internet is an expert at something these days, including me. The internet isn't at fault; it's just made spreading the bullshit easier.

Spotting the bullshit is easier when you realize most people are operating from personal motivations. All people are under the influence. It's just a matter of which ones. Figuring out the main drivers of their grift, conformity, vanity, power seeking, self-promotion, consumerism, extroversion, etc., etc., etc., is like a magic force field for yourself. If you see it for what it is—or see *them* for what they are seeking and what action they are trying to persusade you to take—it's easier to step out of sync with the norm. You no longer have to fall under the spells of conformity, likability, and authority.

You can do the most courageous thing of all. Be weird. Be you.

Embracing your weirdness, the very things that make you one of a kind, is an easy tool to release the fear of not belonging. That deep-seated trepidation, a basic human instinct, drives us to act outside of our values, turn from what we know is right for us, and dampen our inner lights. To overcome that and be okay with letting the chips fall where they may by being gloriously and unapologetically unique, we can turn to, and honor, the inner weirdo. When you find yourself sharing the same beliefs, thoughts, likes, dislikes, styles, and looks, take a moment to see if your differences have been cast aside. Daringly seek out your distinctions and let that freak flag fly. Bring to light who you really are and become comfortable with embracing yourself.

I'm desperately trying to counterinfluence the world for both Hays and now my younger son too (while also teaching them not to tell an adult that their job is evil). Because while there are other ways to manipulate, persuade, and, most important, navigate the world, maybe the greatest trick is in the permission to become the outlier. I so badly want them to *not* belong.

In my family, we have a saying that has been passed down through the generations, restated just incorrectly enough again and again, as if in a game of telephone, that we can no longer determine the original language. But the gist of it is this:

Everyone is weird except for me and thee, and I'm beginning to wonder about thee.

We say it often. About everyone. Mostly about ourselves.

Because family *folklore* and the stories we tell ourselves inform who we become, I recently added my own postscript: Weird is good.

Folklore   People are interested in *you*.

Folktale   I'm secure and beyond flattery, fear, and conforming.

Folk Rebellion   I know that deep down people just want to belong, and that may cause them to do things outside of their character. But that doesn't mean that I have to.

__Raise Hell__ If you were to begin embracing your inner weirdo, what would be the first things you would cast aside? Inane small talk? Business casual? Swallowing your thoughts?

Choose a day to do the unexpected. If everyone is going for a walk, stay inside and read. If a group email has everyone responding, see what happens if you don't. When it's dinnertime, opt out and take a bath instead. How do you feel when you don't conform in these moments? Survivable? Great. You now have the secret to breaking free from conformity's clutch.

Life gives you a lot of chances to screw up, which means you have just as many chances to get it right.

—*Sex and the City*

Part III
The Bad Influence

Now that you know *where* influence lurks (Part I) and *how* it works (Part II), here is *why* influence might be wielded—and why, even though you might be tempted to brandish it yourself, you don't want to be seduced by it!

Motivations.

If we were to look at my time as a ballroom dance instructor (influencer) through the prism of motivations, we would be able to see that they were flowing into and out of me like a suncatcher. Let's examine this example more deeply.

In general, motivations are threefold: (1) What the influencer's intended outcome is on the influencee: a change in behavior, a change in belief, a change in attitude. (2) The influencer's own motivations for desiring the change in a behavior, a belief, or an attitude in someone else. (3) And the underlying motivations of the influences that the influencer is under.

I know. Have your eyes crossed yet? Let's break it down.

1) My intended outcome was to change a behavior in my students (the influencee). I wanted them to buy lessons.
2) My own motivations for desiring that change in my student's behavior were money, accolades, and ambition.
3) The influences I was under and *their motivations* were many but here are a few—my boss was motivated to become a "top studio," my parents desired me to "find a real job," pop culture inspired me to take advantage of the cultural zeitgeist, the societal norm that achievement should be met at all costs convinced me to pursue achievement too, and, of course, capitalism influenced me to continually buy things and pay off the debt that followed.

A person's motivations can be as simple as wanting to make a purchase and can be as complicated as psychological factors driving them to become a global leader!

How strong the motivations are, *their why,* is often indicative of why an act of persuasion may go from being seductive to persuasive, decisive to coercive. The influencer has the choice in making their attempts to influence, weak to strong, implicit to explicit. When the stakes are high, often the attempts to influence become more questionable. If I was short on rent or the studio was close to hitting a goal, my arsenal of tricks and tactics usually increased from being subtle to questionably aggressive. It might be said that persuasion can have the best interests of another in mind. But in contrast to persuasion, having only your own best interest at heart might lean more toward manipulation. It's like walking onto a car lot and being at the mercy of the motivations of the salesperson on that day. Is there a goal to hit, a specific car to sell, a job on the line, a superior to please, or mouths to feed? If the motivation to buy is what gets you onto the car lot, it's influence that points you in a particular direction of a certain car.

There are three main intended outcomes in influence: a change in behavior, a change in belief, or a change in attitude. Inducing a change in behavior is called compliance. Inducing a change in attitude is called persuasion. Inducing a change in belief is called either education or propaganda—depending on your perspective, and *theirs*. When you successfully persuade someone to change their mind, it is rarely a result of pleading or threats. Instead, most successful endeavors in influencing a change necessitate thoughtful planning, argument structuring, the presentation of concrete facts, and an attempt to connect on an emotional level with the target. And in places where an incentive is too irresistible, something that cannot be refused because of need or something else, it tips out of persuasion and into coercion.

As you read through the following pages, you'll see not only the *why* for me, but the *why* of business entities, policy makers, public relations, and leadership. Why I, we, and they want to persuade, educate, or force compliance. The motivations are many. Extrinsic ones have more to do with the desires outside of oneself, such as power, publicity, fame, money, or status. Intrinsic motivations have more to do with growth, fulfillment, learning, meaning, or purpose. In this section, I will share some of the times I maliciously used my powers of influence as I fell more and more into the roles of an influencer, self-promoter, salesperson, brand strategist, communications executive, and thought leader, and what drove me to take those actions. You'll get a clear, unencumbered, no-holds-barred look behind the curtain of influencers—all the bad and the ugly included.

And while there is power in knowledge, as you will clearly see in the pages that follow, you must tread lightly, as this great "seeing" can ignite an alter ego, or just plain ego. In your quest to not be influenced by societal norms, you may instead want to become an influencer—a salesperson, trendsetter, advertiser, brander—just as I did. We all have goals and dreams; we all have motivations; and we are all under the influence of cultural, personal, social, and psychological factors of our own (*Inner, Surface,* and *Outer World*)—so the tricks of the trade I'll open up about may look enticing as a means to an end for some of your own missions in life. But please don't be tempted to give in to using them just yet! Let me show you all the ways in which I practiced being the bad influence over the course of my career, a journey from bartender to brand strategist to *Folk Rebellion.* I'll come clean with how I influenced behaviors, beliefs, and attitudes. I was all balled up in both intrinsic and extrinsic motivations, and even at times, manipulative—and I'll explain why you don't want to be seduced into doing the same. Acting as the influencer completely upended my life, and I'll show you why you'll want to avoid following in my footsteps (albeit fun at times!). In place of giving in to the tug of the "bad influence," I'll show you how to take advantage of and embrace the much more rewarding alternatives: *real pleasure, reaching higher, building friendships, nurturing interests,* and *living purposefully.*

PONY UP

(or Embracing Real Pleasure
Instead of Short-Lived
Dopamine
Hits)

The most powerful position I've ever held was behind three feet of pine.

I wouldn't call it glamorous, per se. Each night, I waded through foul, grimy mudwater—a mixture of spilled booze, beer foam runoff, spit, and spray from the soda gun we used to put out the flames I lit atop the bar. But when I held court over a crowd desperate for permission to have a damn good time, my fellow bartenders and I provided it, nightly, with reckless abandon.

The Red Rock West Saloon on the corner of West 17th Street and Tenth Avenue, on the cusp between New York City's not yet gentrified (at the time) Meatpacking District and Chelsea neighborhoods, had a weathered and stickered door barricaded by red velvet ropes and a giant biker as bouncer. If you had the guts to step up and through that door into this den of iniquity, you were also busting through the monotony of life—even if only for one night.

In there, you could experience a kind of uninhibited, visceral release. You could go feral. We had mastered and doled out the intoxicating cocktail of sex, drugs, and rock 'n' roll to trigger pleasure. And we wielded that influence without shame, targeting our customers—and definitely their wallets. We weren't good for their health or their bank accounts—we were good for a damn good time.

The human desire for pleasurable experiences, like the one we were providing, is hardwired. And in some ways, and sometimes, it is good for your mental health to let loose. Dopamine, the brain's feel-good chemical, gets released while you're having that damn good time. It's the neurotransmitter that comes from feeling arousal and joy, which is exactly why we chase it. And it's not just connected to a raucous night out. When we listen to good music, have sex, smell cookies baking, take certain drugs, smoke cigarettes, go shopping, create art, eat delicious food, gamble, or even scroll social media, our body experiences the dopamine rush. Of course some manners of achieving those hits are better for our overall health than others.

Believe it or not, getting a hug, eating candy, and snorting cocaine all release dopamine in the same way to make you feel good. Once we've had that pleasurable experience, the brain is a quick learner and wants to replicate that feeling again, which leads us to seek out and repeat that reward. Scientists refer to it as the "teaching chemical" because it teaches us to go back for more. That's why pleasure is one of the big influences over our behaviors, motivations, and emotions. And like all things influential, it cuts both ways.

In looking back, I can see how we were dopamine dealers behind that pine—and that gave us crazy power. We'd joke that the physical bar was there to keep the riffraff away, though we were clearly riffraff ourselves. We loved that godforsaken hellhole—some kind of *Animal House* meets *Cocktails* meets *Hustler* explosion.

The first time I stepped through the door of the Red Rock West Saloon, I, too, fell prey to one of its masters of influence—bottled, shaken, and poured into a pair of vinyl pants. Her name was Sunny.

It was just weeks after 9/11 (and years before I found my dream apartment with the cherry tree out back, during my zip-code-hopping phase), and I'd invited my upstate friends to New York City for one last go-around before I tucked my tail between my legs and moved home. I'd moved to NYC only three weeks before to try to sell a TV idea to MTV, a partner-dancing show that no one could see working (don't even ask me about *Dancing with the Stars*). But the tragedy of 9/11 had upended our entire world and was making the search for a side gig impossible. The city was done with me, it seemed. My swan song gathering was to be at that famed wild honky-tonk recently immortalized on film, *Coyote Ugly*. But while we were sitting on the F train, reading aloud from our paper Zagat guide, another passenger interrupted us to say that *Coyote* was second to another rowdy bar, beloved by "true New Yorkers." Aghast at possibly being mistaken for tourists, we changed our plans immediately.

We waited in line for almost two hours. Biker John, as I would later come to know him, finally nodded to us, and opened the door to utter debauchery. Hundreds of sweaty bodies, butts to nuts, were packed in like sardines, screaming, singing, snogging. A megaphone roared with a female voice that sounded a thousand cigarettes deep asking the raucous crowd: "Are you having a good fucking time?!" I looked up and spotted its source atop the bar.

When the crowd didn't respond loudly enough, she asked again, this time at full volume—while lighting her fingers on fire and blowing flames out of her mouth. My friends knew I wasn't a fan of crowds, super loud places, or feeling trapped, so they pushed me on while I tried to retreat, slamming me through the crowd until I hit the bar and came face-to-face with Sunny.

Sunny, a veteran of Red Rock, had seen enough timid, claustrophobic party-poopers to know by the look on my face that I was getting ready to bail.

I'd later learn that a gaggle of women was like chum for bartenders. She needed us (fresh meat) to stay to attract the guys with unlimited corporate cards (fish in a bucket) so she could start a tab and ring it high (catch and kill).

Agile and strong, in seconds she had pulled me to standing on top of the bar, for the first time in my life. "It's not crowded up here!" she said. "You stay here all night, and I'll have the men buy your drinks. I'm going to make so much money off your ass!" She laughed, slapping my butt, and hopped down to find her big fish to fund us both. There was no disobeying Sunny. That much was clear.

Under her (questionable) influence, we ended up having the time of our lives that night, followed by the hangover to match the next day— and surprisingly, a job offer to boot. I didn't know it yet, but I was about to become the protégé of another master of influence. Within twenty-four hours, I went from ready to slink back home to Mom to a whiskey-slinging bartender at one of NYC's most debaucherous watering holes. I was instructed by the owner to "go see Jimmy at Trash and Vaudeville on St. Marks for clothes." At Red Rock, jeans and T-shirts wouldn't cut it. Anyone could wear jeans. We were unique, captivating, projecting power and authority, and our clothing needed to signal that.

I'd later learn that Jimmy was East Village royalty, friend to legends including Debbie Harry, Iggy Pop, and Guns N' Roses. The face of the infamous rock 'n' roll and bondage clothing store—and my stylist—had toothpick-size legs poured into low-rise leather pants and wore a vest with no shirt. Jimmy and his bottle-blond hair greeted me cheerfully, leading me by the hand to a fitting room. He loved when the "Red Rock girls" came in, he said. Flitting about the store, he scooped up the most ridiculous things you could call clothes that I had ever seen. He fitted me for a pair of vinyl pants with blazing stars down the side, leather chaps to wear over cartoon Underoos, fishnet arm sleeves, a corset with ceramic bunny heads on the tits, towering platform boots ("Functional but fucking rock 'n' roll!" he exclaimed), and other ripped, shiny, or see-through pieces. Like a punk *Pretty Woman,* he told me he'd put it on the bar's tab. I'd pay it back later with tips. The total was more than my rent.

Red Rock training was a trial-by-fire situation, which was how they weeded out the ones who wouldn't go the distance. They told me to show up Friday night at nine P.M. to "learn the ropes" during the busiest hour of the week. Mostly, they wanted to see how the crowd liked me, how I handled my shit while hundreds of people shouted at me for drinks, and how I handled my liquor. We were expected to drink, and drink heavily, with every person in the bar. Quite literally, it was my job to be under the influence along with my customers—functionally hammered all the time.

But not Sunny. Sunny *never* drank. Impervious to influence on some degree—she was the only exception. This didn't stop her hustling, though. It took me months to figure out that when she had customers buying rounds of top-shelf vodka, she was spitting hers out on the ground during a sexy (distracting) hair flip. And when a customer started to become aware of her schtick, she would play dumb and move to spitting it between her cobartenders' breasts—much to our dismay. But the distraction of this form of motorboating worked, and she could continue to sell hundreds of dollars' worth of shots and remain sober all night. She was always in control.

In the same way my training at the dance studio was more about sales than dance, learning to bartend included virtually nothing about making actual drinks. To this day, if you ask me for a margarita, you'll get a shot of tequila with a lime, just as Sunny taught me. What I was being taught instead was how to ring the register, how to maximize the credit card tabs, and how to keep the customers hooked so they'd stay longer. Along with learning how to maximize profits, I also had to develop my schtick. What was the role I would play to add to the mystique of our shanty-cabaret dive bar? I wasn't the toughest, the meanest, the prettiest, or the sexiest. So against the odds, I chose the nicest.

That choice of persona worked to my advantage. First of all, playing the "aw shucks, who, me?" girl next door allowed me to avoid the more salacious things I didn't want to participate in. And it only exacerbated the draw for men, who saw me as someone unattainable they could never actually date, fuck, suck, take home, impress, or make angry.

To be clear, my other bartender mates didn't do those things either, but they let you *think* they would. Their allure was in the possibility; mine was in the challenge. It allowed me to politely decline a multitude of bizarro offers: to be a cast member on *Temptation Island,* the third wheel for a wealthy married couple, or the naked sushi bar for a movie premiere at the club next door. I hid happily behind my facade of "just off the farm" naivety and innocence. And it worked. I could use that sense of inaccessibility to influence my customers into spending more and trying harder. The more I demurred, the more they reveled in the pleasure of the night and sought out that extra dopamine hit of my split attention. My tips would go up as clients doubled down on their propositions. The more I said no, the more they laid down cash.

I learned a new language, a litany of words that meant little to me before—*86'ed, kitty, buyback, comp, ring, last call.* To 86 something meant it was gone. You could 86 a customer or a drink. "86 Bud Light!" would be shouted in the same vein as "He's 86'ed!"—a directive to the bouncer to throw some guy's ass out. A *kitty* was the pile of money we trained our customers to ante up on the bar in front of them. Instead of reaching into their wallet for each drink, our regulars would just leave a stack of cash there in front of them, which would make it quicker for us to tally and serve. Everyone would throw in money, which we would pull from all night, letting them know when it was running low. But the psychology behind it was the biggest manipulation: No one wanted to look cheap or uncool, so no one refused to throw in. And at the end of the night, in order to appear important, often they'd just tell us to keep whatever was left. The kitty made us lots of cash, often more than $1,200 a shift, which was always the primary motivator. A *buyback* was what we called the "Hey, this one's on me!" free drink we sometimes gave our customers. As a rule of thumb for our bar, a freebie was offered after every third paid-for drink, but only for regulars or first timers and never with the top-shelf stuff or for bachelor/bachelorette parties. A *comp,* which stood for complimentary, was our monopoly money. If the owner came in, or a coworker, or someone who merited a free bar tab (our tattoo artist, hair-extension stylist, family, broke roommates, or someone we got drunk enough to forget to pay their bill on the way out the door, which was considered our fault for doing our job too well), we had this special tab. In general, the value of the comp tab was always mysterious and up to our discretion. There was no set number of comps, but the barometer was that it should always be in relation to the *ring.* Your ring was what they expected you to accumulate on your given register, depending on the shift and day. A Friday-night ring was higher than, say, a Tuesday afternoon, but the goals were equally lofty. If you missed your ring often enough, you'd lose your shift to someone who would make sure they didn't. In other words, if you wanted to keep your job, you had to know how to wield your influence.

In much the same way tech, casino, and online gambling executives figured out how to pull the triggers of dopamine pleasure with triple-seven slot machines, we hooked our clients with jackpots of endless depravity, set to a soundtrack of Willy, Axl, and Johnny blasting from the old juke.

Objectively, the place was a dirtbag-ridden, divey mess and, against that backdrop, we seemed brazen, unfettered, not of the real world. But that was the whole allure. Slumming it with us meant giving in to curiosity and adventure. Being low-rent for a night had its upside.

"The Rock" was an escape from the job, the wife, the boyfriend who didn't pay attention, the laundry, the news, the keeping-your-shit-together of life. In there, you were meant to lose your shit. This was escapism. It didn't matter if you were a banker, a school nurse, a fireman, a dropout, a drug addict, or a mom. We treated everyone the same. This, we understood, was the intoxication of the place. If we could convince people that the typical rules didn't exist, that they were safe to let loose, they'd stay as long as they could, and we'd make unconscionably excessive tips. Such was our influence over them.

The first time I saw Sunny cut off a man's ponytail, I was floored. How could she do that so nonchalantly? How could he still be smiling? Once again, we can probably blame our brains for that. According to an article in the *Harvard Business Review*, "How Hardwired Is Human Behavior?," our most innate instincts are rooted in our reptilian brain, wired for survival first and foremost. We think: *Can I eat it? Can I have sex with it? Will it kill me?* Back in our caveman days, this was what we needed to make sure the species survived. The dopamine neurotransmitter is reptilian too. These primitive parts of the brain keep us in a constant battle between desire, responsibility, and the ways in which our modern world is working to exploit it.

Sunny was confident that the ponytail victim was operating from his reptile brain—aka his desire to have sex with her. She knew he'd get a thrill out of being dominated by a woman he thought was hot and that, even in chopping off his hair, he'd be more likely to reward her with tips. Just like Sunny exploited his dopamine rush using booze and tits in his face, there are everyday distractions in today's world designed to tap into and exploit our dopamine receptors. The problem is not only that we get hooked on those bursts of dopamine, but also that what goes up, must come down. Scrolling through our phones, getting likes on social media, and watching TV all flood us with happy chemicals—at first. But the more addicted we become, the less joy they bring. When the likes on our Instagram posts slow down, we find ourselves in deeper despair than before we ever even looked at our phones.

The brain doesn't know the difference between actual joy and dopamine addiction. Because dopamine isn't directly responsible for feelings of euphoria or pleasure—remember, it's the *response* to your experience of pleasure—it doesn't have much to do with *creating* actual pleasurable feelings. It's a dangerous thing when our bodies and minds don't connect and work in sync to comprehend the reality of a situation. The levers of dopamine desire can be pulled without legitimate pleasure and, often, in its place sit addiction, loneliness, listlessness, emptiness, sadness, anxiety, or depression. When we keep doing something (playing a video game, for example), and we aren't sure why we can't stop, though we notice it's not bringing us happiness, that's what's going on. Red Rock was mind-blowing, liberating fun in the moment, but, the next day, that ponytail guy was going to wake up with a giant credit card bill, a killer hangover, and short, cleavered hair, only to realize that he didn't even get any action. In some ways, the digital world is more dangerous because there's no last call, no blinding white light of reality to reveal the ugliness hidden in the dark.

Desire comes in all shapes, sizes, and forms, and at all ages. People in the hospitality industry will tell you drunk adults are just like toddlers. They want what they want, and they want it now. They laugh with abandon, wobble when they walk, and often, as the night wears on and they get tired—they cry. They are dysregulated in the same way a child who hasn't learned to control his emotions is. This is why booze is used to lubricate business dinners and dates. For one side, it provides liquid courage; for the other, convenient susceptibility to influence. When your walls are down, so are your defenses, allowing you to become more easily manipulated. For us bartenders, we knew the quicker we got you lit, the less likely it was that you would leave, for just this reason.

Our customers were clued in to the fact that they were doing something that gave them a high. Yes, we influenced them from the moment they walked in the door—from manipulating their state of mind to creating a feeling of connection by drunken barroom sing-alongs to old favorites like "Friends in Low Places." We hustled their credit cards and lit a fire of freedom and fun in their minds by lighting the bar, frying pan, or our fingers on fire—but that all ended at four A.M. And even if there were points of the night that were foggy, in general, people knew what was happening to them. It's why they came: to get lost for a night, to forget, to be a different version of themselves.

Today our dopamine is being preyed upon but without our really knowing it. I'm not excusing our bad behavior back in the day. Through my lens today, I can see the trouble in the environment we created: our self-objectification for money, the reliance on substances to modify behaviors, the possible danger we put our customers and ourselves in. I was a bad influence, a wheeler-dealer manipulator, even if the transaction between me and my target was, in general, explicit. This place was a relic of its time. And while I'm glad my stint as a devious dopamine dealer is in the past, I was glad to have been a part of the era—both for what it taught me and for the sheer experience of it and the lifelong friends I made (Stevie! Tracy!). We have the internet for viewing wild shenanigans now, where being triggered for pleasure is less overt and in your face (fire and tits), and more imperceptible and unseen (more like the air we breathe).

Today targeting dopamine covertly is big business.

Back then, each of us in our role—owner, bartender, shopkeeper, customer—were trading X for Y. I'm not so sure today's digital universe and all its manipulative dopamine dealers work that way. They live so silently in the shadows that before you know it, you've destroyed your relationship, health, job, and life unknowingly. Sure, people had to be wary of us in those days. I was out there behaving in ways I wouldn't have necessarily imagined previously, even when I found the antics to be questionable. Why? The ego boost. The sense of power. Most of all, the money. But we did give them something back. When I think about my gory, glory days, there was a spirit of connection that at least lived inside those transactions, one I long for from time to time because today's transactions don't hold that same spirit. In our free market, that loss is only going to keep growing. Cravings come in all different forms. A lack of impulse control isn't attributed only to drowning yourself in a bottle of booze. Today's insidious influence of dopamine drivers looks more like overexercising, obsessive self-care rituals, or an overreliance on online relations instead of IRL ones. Tits included.

So as the platitudes say on the ironically unhappy-hormone slot machine we call social media, to not just survive *but thrive!* in our modern society, we need to become cognizant of the negative influences targeting our caveman biology. What's being served to us, and our reptilian instincts, has become progressively dominant. It's everywhere, more invasive, faster, and shows no signs of stopping. Turning away from what's bad for us, but feels biologically like adulterated fun, requires a shift in power. The customers of Red Rock held all the leverage; they just didn't know it. Their attendance, money, and opinions could make or break us.

If we can identify when we're just getting sparks based on how we feel afterward, maybe we can start to see how we're being unscrupulously triggered. Dopamine-fueled rapid rewards, stimulation-driven environments, lack of impulse control, a new experience each time, and consequence-free access all leave us feeling less fulfilled than the more meaningful and human deeper dopamine-driven pleasures our biology was built for. Then, maybe, we can choose not to partake in the businesses that benefit off creating or feeding our addictions and, thus, release certain behaviors that aren't serving us. I'm talking about the experiences that offer only the briefest spark, like turning to the repeated ease and convenience of pornography until you've turned away from your partner with no turning back. Or optimization of body and health through constant tracking and monitoring that leads to six-pack abs and a neurosis of navel-gazing busyness through data that furthers feelings of disappointment, unattainable standards, and obsession with self.

Instead, we can turn away not only from being the bad influence, but also from the dopamine dealers who light us up superficially before we dim like a burned-out neon sign—low voltage, shorted out, and broken. We can stop settling for short-lived sparks and instead turn toward the things that light us up long term—connecting, learning, loving, playing, building, growing, creating. And that way, and that way alone, maybe we can hold on to our ponytails.

Folklore   Pleasure equals happiness.

Folktale   In order to be happy, I need to nurture pleasurable experiences.

Folk Rebellion   I know that all pleasure is not created equal, and my biology can be used against me for the benefit of others.

Raise Hell Weigh the difference between the things that give you short-lived dopamine hits instead of lasting pleasure. Consider how they look and feel different when they're delivered online instead of offline. What is the feeling afterward? Are you depleted or energized? Are you able to walk away, or do you keep coming back? Do the things that trigger your dopamine trade in the currency of connection, or are they leaving you disconnected?

GOOD ON PAPER

(or Reaching Up Instead of Falling for Fakes)

I've faked my way into a lot of things: Good moods. Bars before I was of age. Weddings where I wasn't a guest. Concerts for which I didn't have tickets. Box seat suites when I was a bleacher creature. Bravery in the face of terror. Relationships that were all wrong. Celebrity parties, classes that were full, and clubs with giant bouncers who wouldn't let me cut the line. But of all the things I've faked my way into, my career is the most outlandish.

"Fake it till you make it" is well-worn advice for a reason. The truth: It works. If you think about it more generally, it basically suggests that we should behave as if we're confident, competent, and optimistic even when we're not. The hope is that faking it this way will help us manifest the qualities we're mimicking into actual existence. The result, if all goes well, brings us one rung higher, one step closer. The adage is thrown out as an encouraging kick in the pants when there are no right words for coping with a difficult personal situation that won't follow the reason or logic of time (like a broken heart, for instance). It's for courage when a new promotion or opportunity feels like biting off more than you can chew. It's a remedy for imposter syndrome.

There are two schools of thought for this bit of life advice: To some, it feels problematic—unearned, irresponsible, unfair. Typically, you hear this complaint from those who earned their success by getting seats at the table in traditional ways—through years of study, college relationships, recommendations, networking, or connections that open doors to coveted internships and, eventually, high-level jobs. For others, faking it is seen as the necessary scaffolding built up around workarounds, self-education, autodidacticism, mustered enthusiasm, and hands-on experience. For those who collect trophies along the road to success in more standard ways, as if ticking off boxes on a checklist, confidence and optimism may abound more naturally. For those who don't have an automatic seat at the table or who are less inclined to function easily within accepted structures, confidence and optimism must often be postured before they're internalized.

At twenty-two years old, I found myself in a catch-22. I wanted to beef up my résumé with more "real" job experience, but I had no college degree, which was often a prerequisite for getting some. So I decided to try to become a guest professor at a local business school the Jess way, teaching adults much older than me how to build their small businesses in the digital world. I simply developed a course outline and pitched the idea. Was I a trained professor? No. Did I have a degree? Didn't come up. Why not? I knew I had something else they needed. The business school wanted to appear ahead of the curve in terms of innovation, tech, and social media back in the midaughts. At that time, I was using Facebook for marketing local wine events that I was throwing as a side hustle and selling online ad sales for a local newspaper. To the school, I was a digital virtuoso. To me, the school was a ticket to legitimacy.

They agreed right away. And that experience of teaching Viral Networking 101, a class I made up out of thin air, was the gateway to my future gig, a cushy corporate job with a "real" job title: digital strategist. Just like that, an entire new career path was born.

It starts innocently enough. You appear overconfident in interviews. Dress the part. Identify something remotely similar in your past and find a way to align it with the opportunities presented. Use outward signals and displays like clothes, body language, grit, tenacity, gumption, and guts to create a first impression of ability, and *then* inward change follows.

Despite my lack of specific education, I delivered on my course offering, gaining private clients from the experience, and an invitation to become a board member of a local cultural institution. I was officially validated. With this key step, I not only helped influence how the school was positioned in the market and how the students and their businesses were perceived, but I also influenced how people perceived *me*. Finally, I was good on paper.

So sometimes you need to fake it till you make it. That's what I believe. God knows I did. I may not always be proud of the tactics I used to play catch-up or boost myself onto a somewhat more even playing field (social mobility), but by doing so, I've given myself so many other things to be proud of—a career, a voice, a platform, a family, and all the accolades that I earned once I got in the ring. Of course none of this can happen without ingenuity in advance and hard work once you're given an opportunity. I'll take that credit. And I'm not really apologizing because faking it is a skill unto itself, and my ability to fudge my way up the ladder is now useful to you as a kind of warning. Sure, many people fake it till they make it with harmless intentions to themselves and others. But for fuck's sake, watch out for people who are just faking it—full stop. It's easier than you think.

There's been an increase in those types as of late in our culture because the internet makes it easier to pretend. What were once innocent Stuart Smalley–style positive affirmations—"I'm good enough, I'm smart enough, and doggone it, people like me!"—paired with exaggerated smiles, overly firm handshakes, and truth-stretching in order to make connections, have turned small affirmations into big problems. The power of technology mixed with social media and our digitally altered universe has given rise to everyone becoming their own personal brand or commodity with a platform to sell a fake feeling, talent, service, festival, business, expertise—life. Today all a faker needs is a ring light, ChatGPT, or an @ handle.

In recent years, there have been periods in which certain professions became rife with fakers, both benign and nefarious. I call it "speculating." I love a job that requires little-to-no red tape. I've had many. They are some of the best roles for getting real world experience, which is invaluable when you don't have any, or don't have the right kind. But at certain times, specifically potentially prosperous economic times, these career shifts tend to draw more people with more self-serving interests than just a love of real estate or wanting a fresh start. Before the housing market collapsed in 2008, everyone and their mother (me included) was getting real estate licenses because there was so much money to be made. You didn't need to do much to become an agent, you just had to fake it till you made it. Just pass a test, then pay into what was essentially the pyramid scheme of your agency and team. I happened to be living in Florida at the time (on one of my many breaks from NYC), and if you weren't an agent, you were being encouraged by even the mailman to "get your license!" You made no money at first but got a title and business cards and began cold-calling clients in hopes of the windfall. Maybe you got training, or you trained yourself. Either way, you faked the expertise until you had some. And while there are plenty of responsible, respectable, educated experts in real estate (both from on-the-job learning and through school), at that time, the influence of opportunity outweighed the expertise. How could an inexperienced and largely untrained broker foresee, understand, or explain to a house hunter why selling subprime mortgages with zero dollars down for the purchase of three condos at once could kick-start an economic depression, devastating thousands upon thousands of lives? And why would people under the influence of wanting to make money, a living, and a name for themselves tell anyone otherwise? The layers of people under the influence were many, all of them influencing one another: agents, brokerages, inspectors, builders, mortgage lenders, real estate developers, banks, insurance companies, and on and on and on until *pop!* This phenomenon was in part (along with a lot of other factors) to blame for the horrific way the housing bubble burst. The stakes when fakers put themselves first and use their influence only to self-serve can be quite high; many lost their homes to foreclosure when the housing bubble burst because they hadn't received good, meaningful, well-intentioned guidance. With great influence comes great responsibility.

Now life coaches are the new real estate agents. They pop up on your screens as you scroll, offering you unsolicited advice in a calm, even-keeled, tranquil way. They have glowing skin from their "good light" and weirdly glowing eyes from their ring light, and deliver their messages in easily digestible, Instagrammable sound bites that the Explore Page algorithm loves. *They're* your friend, your neighbor, and, yes, your onetime real estate agent pursuing a new path. And most likely, they have no business giving you advice, especially when you didn't ask for it. But now they're under the influence of opportunity, social media algorithm shifts, and the zeitgeist of the times. They're embodying the profession in dress, presentation style, and lingo (same as I did when I bought my first pencil skirt, put on classical music, held a gold-plated clipboard and said things like "Welcome! Mr. and Mrs. so-and-so," shaking hands feverishly during my first open house). They're wearing glasses and taking deep breaths and pointing to the text over their image encouraging you to "find your truth," or "overcome your neurodivergency in love," or "raise mindful children," or "find your joy through microdosing." And these people don't need licenses to show up on your screens. My concern is what influences these experts are under, and what they are guiding you toward. The stakes are even higher when they're advising on relationships, nutrition, or mental health—if not authentic, ego-motivated advice from these fake-it-till-you-make-it gurus could be downright life-threatening to those who follow their suggestions without question.

In the gold rush of misery, despair, declining mental health, pandemic poverty, and midlife (or even quarter-life) crisis, these new speculators are only following the trends of opportunity. Who can blame them? We're all so fucked up and desperate for relief that we're willing to try anything. And in a terrible twist, *anyone* can become a supposed expert in unfucking up or recalibrating yourself, your kids, your plants, your pets, your grief, your closets, your diet, your productivity, your screens, your relationship, your style, your personality, your purpose, your finances, your health, your business, your side hustle, your art, your writing, your voice, your truth, your trauma, your podcast, your future, your worth, your hopes, and your dreams. And while some stakes are less kill or cure (plants and messy closets), others are very high (money, relationships, mental health). Today everyone is a guru of something.

Back in the day, when I taught my class, people didn't come to me to save their lives, relationships, or find their purpose. I'm not saying my exaggeration wasn't self-serving—it was. But I was teaching people how to set up a Facebook page—*a Facebook page*. And that was something I really did know how to do, whether I was technically trained or not. It was low stakes. The stakes are higher when people who want to fix their lives, their marriages, or their futures look outside of themselves to any person claiming to be able to help. And while on-the-job training through learned experience is something I very much believe in, people offering high-stakes life advice shouldn't be learning their craft while on the job.

There are a million tools to explore for self-discovery—well-researched books, journaling prompts, self-assessment tests, therapy sessions with an actual trained professional, or even simply taking some alone time. Of course we can all understand why it's comforting to have a guide to help you get out of your own way, your own head. I've had a bunch of collaborators, friends, and even hired helpers along finding my own way. But the question that you must ask, even at your most desperate, is: What are the intentions of the person you're paying?

As we've established, the most successful person isn't necessarily the best. The next big fitness guru, wellness thought leader, productivity expert, food movement podcaster, or feminism-for-sale icon can be plucked from obscurity on TikTok or *Real Housewives* or be seated beside Oprah before followers can even "click, like, subscribe." Who is choosing which "experts" should be legitimized, featured in trusted magazines and on talk shows? Is it someone who cares about your well-being or just a studio executive trying to cash in on audience numbers? Is it because this "expert" is so incredible at helping people or is it because he, she, or they are telegenic and charming?

This kind of overnight success, found without any particular credentials or ability, is a big carrot now, possible for anyone anywhere to eat—if they just chase it enough. Being the poster child for a movement is massively lucrative to the guru and the brands attached to them. Becoming the next big thing in self-help equals earning outrageous speaking fees, selling private coaching packages, and participating in media tours to sell the empire. And the empire can include anything and everything, once it's been established: co-working spaces; brand endorsement deals; TV shows; and "empowering" product lines of beauty regimens, fragrances and candles, notebooks, yoga wear, wellness drinks, minimalist kitchen instruments, and more.

For some gurus, a book is just an extension of their brand, a rung on the ladder to fame and fortune. Every person they meet is a potential sale into their pipeline or funnel. My problem isn't the entrepreneurship itself, but the self-serving motivations behind it—at the expense of people's livelihoods and mental health. It's not that all people with ambition are bad; it's about the motivations driving them. These people are performing either to get rich, get noticed, or satisfy the constant voice in their head insisting, "If you don't get attention, you don't exist." Those are the influences that make them turn around and try to win your attention, shouting, "Hey, look at me! Look how great I am! I am winning at life! I am fine. Everything's fine if you just do what I say!" Lightning-fast carrots, just out of reach. Bigger stages, bigger stakes. I'm reminded of the idiom that says those that can't do, teach. Sometimes the people teaching others need these lessons the most.

The biggest concern is that these "gurus" are attempting to help individuals feel better, be better, live better. The problem here is the pervasive idea today that we all *need* to be better. And if we actually *do need* to get better, for some reason, then why are these solutions only targeting the symptoms instead of the causes? When we rest the responsibility for feeling better (as if being content all the time is even an option) on the shoulders of individuals and tell them that if only they sign up for this $399 course, they'll stop feeling sick, tired, lost, uninspired, alone, angry, etc., we are becoming a product of the illnesses that really ail us. If we really feel this way, ironically it's most likely because we're constantly being targeted this way, that the perfect life is just within our reach if only we could be a little bit better.

The solution to true fulfillment is not in content from a professional organizer or the purchase of joy-sparking spatulas. The actual solution is in the realization that our society is set up to make us feel sick or inadequate and then throw money at the discomfort so others can profit from it. But you, dear reader, have unlocked how to outsmart them and not be sucked in. You know to ask, first, "What are their intentions?"

Stop telling us we can feel better if we buy certain objects, eat certain foods, fill our house with plants, learn how to manage our unhappiness through these daily practices sent via email to our inbox of shit. If we feel bad, it's because we're constantly being told that we should, and no amount of consumerism is going to solve that.

I know this firsthand. For a period of time, when, at first, I wrongly believed that I'd had an epiphany and had ducked out from beneath society's thumb by unplugging, I worked in this realm of guruness. After a while, it became easier to talk about my work than actually do the work. There was no joy in it. I was once again a cog in the machine en route to prosperity. I needed to walk the walk and talk the talk. Practice what I preached instead of performing.

This book is not my calling card, a step toward endorsement deals and a product line. I've done all that. And during that time, I thankfully learned that all I wanted to do was be creative, make things I am proud of, and communicate to others. My bio now says "storyteller" in the lead. Capitalism has tried to consume that too, making the word a more nuanced way to say "selling." I think that means we need true storytellers, now more than ever, to reclaim what is so vital to our human experience. We need to expose the smoke and mirrors. So after many decades of learned experience (in life, writing, and influencing), I have the pleasure of writing this book, sharing a story, with zero faking.

Folklore All successful, notable, self-promotional people are experts.

Folktale I can trust this person, and I must get a degree to be like them.

Folk Rebellion I must do my due diligence into the background and motivations of people I look up to or want to work with or learn from, trusting my gut to determine whether they have my best interests at heart. Personally, I may not necessarily need a degree, but to become a trustworthy resource I can utilize lived experience, knowledge, and education—not just self-promotion.

*Raise Hell Is there something in your life that you are overcomplicating by outsourcing the knowledge? Are you paying a coach to fit you for a job you know deep down will never work? Is the online community that gives you tips and tricks to care for your plants filling your need for plant tips and tricks or is it actually the community you seek? Are you seeking experts to supply answers to questions you already know but want to avoid? Are you joining groups not because of the content, but because of the need for connection? When you hire a consultant, rep, coach, or vendor, are they the best or just the best at self-promotion? What do they gain by working with you? What are the areas in your life that could truly benefit from well-meaning counseling, and what are some professional, trusted, reliable resources you could turn to and avoid the traps of the fake gurus?

SOMETHING BORROWED

(or Building a Friendships Instead of Favors)

As the infamous adage goes, "It's not what you know; it's who you know." As gross as that might sound in the wrong context, I've found it to be pretty true. Whether you want to score drugs or a *New York Times* write-up, the path is basically one and the same: It's all about your relationships and connections. Getting what you want—in influence and in life—is often lost or gained through the company you keep.

I first learned this as a fifth grader when, because my dad "knew a guy," he was able to secure me highly coveted tickets for the sold-out New Kids on the Block show, upping my social currency and winning me a new cool-girl best friend.

I experienced this memorably again when my best friend and I moved to New York City in the summer of 2001 and went to see about a potential apartment—her mother's friend's mother's friend's place. Broke and jobless (making us less than ideal renters), we arrived at the stoop of the Brooklyn brownstone and buzzed the basement apartment so the owner "could get a look at us." Out came a buxom older woman in a housedress. She sized us up, asking our last names (to see if they sounded Italian enough) and asking, "Cath-o-leek?" in a heavy accent. That was enough for her. No paperwork, no lease—she just handed us the keys to our twenties like it was no big deal. It was the beauty of friends-of-friends at work, shared communities, and relationships. Later I prioritized this tactic in my adult life, strategically focusing on building my network and connections to lean on for the benefit of my clients and the brands I represented.

Whether I was trying to get out of a traffic ticket or secure publicity for one of my clients, simply put, I *always* knew a guy.

It's said that we are the sum of the company we keep. Motivational speaker Jim Rohn once famously asserted, "You are the average of the five people you spend the most time with." Sure, as we've learned, we are greatly influenced by those closest to us. Our moods, ideals, sense of confidence, behaviors, and decisions are all under the influence of those around us. But the impact doesn't stop there.

Not shockingly, influence has a ripple effect outside of our immediate communities. What we do or say tends to affect our friends, friends' friends, and friends' friends' friends. This notion of three degrees of influence was investigated and theorized in the renowned Framingham Heart Study, spearheaded by Nicholas Christakis, a sociologist at Harvard University, and James H. Fowler, a political scientist at the University of California, San Diego. The study's findings demonstrate how a happy person can increase a stranger's happiness and, on the other end of the spectrum, how a bad attitude can reverberate far beyond our immediate circles. It seems shit, or happiness, rolls downhill.

So if the company we keep isn't giving us the inspiration (or moods) we seek, there's a chance that we'll find an opportunity to change that, for better or worse, through more remote connections. Just look at Six Degrees of Kevin Bacon, the concept that all people are only six or fewer social connections away from one another. Tracing our links is not just a game (one popularized in the midnineties around one character actor's prolific film career). It's a strategy for those who want to influence their own lives by influencing others, also known in business as the six handshakes rule. The idea is that, if you need access to something or a certain in, you need only dig deep enough into your relationships to find the right path to success. And, today, searching out that one much-needed mutual friend has become easier than ever with online networks and social media platforms listing our "friends" and "connections" in alphabetical order. The "Hey, you gotta guy?" or "Can you do me a solid . . . ?" hustle is now just a quick keystroke away.

Some people adept at influencing understand this better than others. If, for example, they find that the sum of their network isn't working out how they'd like, they'll try to expand. Known pejoratively as "social climbers," these people are hell-bent on engineering the lives they want by strategically choosing who to be in the room with and do favors for. They'll poach your friends, ones more accomplished or useful than you, hijacking the relationship for their own social gain. They name-drop, only accept invites once they know the guest list, and dress to suggest a certain status, all as a part of the plan to grow their network to increase their standing. At their worst, they resort to candle blowing by metaphorically snuffing out someone else's light— discounting successes, ruining celebrations, ridiculing, condescending conversation, or talking behind another person's back—all to make their own light appear brighter. They believe that by putting someone else down, they're pulled up. It's a rise in comparison, without actually rising at all. And they're willing to do that dirty work. These unscrupulous masters of self-promotion understand that "borrowed interest" turns into "borrowed influence" later down the road.

So how do these manipulators insert themselves into the lives of people who might boost their status? Well, for starters, flattery gets a person everywhere. People love to hear wonderful compliments about themselves, so kissing up is one way. But another, which is perhaps more subtle and, therefore, more devious, is becoming indispensable to someone. Not long ago, when I asked a friend how she came to know a mutual acquaintance, my friend shrugged and said, "I accidentally became friends with her." What did she mean? She meant that this other woman made life easier for her until she was simply part of the fabric of their existence. "She just kept offering to watch my kids and walk my dog," my friend explained. "She'd bring me food and come later to pick up her dishes. That's how we got to know each other and now our kids are friends."

One "coincidence" wasn't lost on me: This mutual acquaintance wanted to send her child to the same school where my friend's kids go and, once they started hanging out, my friend wrote her a much-needed letter of recommendation. "How could I say no?" my friend sighed. My friend had unwittingly made a tacit agreement, which is how this often happens. The influencee says yes to the persuader's bid for friendship, albeit a forced one, having nothing to do with common interests or likability but is instead an obligation after so many friendly gestures. This mutual acquaintance made my friend's life easier, though she never asked for favors, and that was enough for quid pro quo. *Borrowed influence.*

This type of exchange is incredibly powerful. It's why nightclubs pay promoters to bring in models and starlets and then provide them with free bottle service. It's why brands create ambassador networks of aspirational people who appear to epitomize the lifestyle they're marketing with their products. It's why there's a whole economy based on creators online: Their friends, followers, fans, and press will give brands the borrowed interest they seek.

Borrowed influence is on steroids like never before in today's social media world with the advent and adoration of influencers. You might think this is where I shit-talk people on the internet who trade their content for money and free products in hopes of influencing their audience on behalf of brands. Well, you'd be wrong. I have no problems with actual influencers because what they're doing—*influencing*—is literally in their job title. There's no pretense there about their intentions. Their followers know they're being compensated for the borrowed influence of the brands they tout. The more disturbing reality is that despite knowing that they're creating editorial bought and sold by brands, some still emulate them thinking it's real life and not a personal brand. We, the witless consumers, are the ones who have no shame! We know what and why they're hawking, and we happily still consume. That's on us, so I can't hate the influencers. In fact, I appreciate the clarity!

It's the more insidious ways of modifying behavior that are the real problems: the ways in which power and position are gained without the influencee (that's us) ever realizing what's happened, the ways in which I used to influence others to win favors.

Back when I was starting out working with brands, there were no influencers or marketing budgets tied to influence-marketing, so we had to make it happen another way. It might be obvious now, with influencer culture being what it is, but back then, if you wanted to get your product out there, you needed to befriend influential people who might share within their networks. It was just throwing spaghetti at the potential friends-with-benefits wall. So at that time, who was influential to us when "influencers" didn't yet exist? Press, editors, buyers, bloggers, owners of established cool brands, movie producers, "it" kids in the new "it" neighborhood, and even the head bartender at the chicest new cocktail bar with the tightest door. My job was to figure out who were the most impactful types for a particular product or niche and how to get on their good sides in hopes that they would do the bidding of my brands, spreading the word through their networks. To do this, I did a bevy of random things.

First, in order to make contact with these people, I threw events with the singular goal of drawing one particular influential person and photographing them with the product in hopes that my editor "friend" at a magazine would print the image. I would do "desk sides," luring writers and editors with gratis lunch, pedicures, or spa days so we could sit side by side and get to know each other. If I could get them to like me, then maybe I could get them to like my brands and write, talk, and share about us to the people in their universe.

As the internet blossomed, I got even more creative, sending product-themed coconuts in the mail to people I knew would find it funny, in hopes that they'd take pictures to post online. I would rent out hotel pools, throwing soirees with synchronized swimmers, DJs, and endless free booze to attract NYC's art and fashion crowds. I would grease the hands of door guys at the best clubs so I could walk my way past the line, appearing important to whatever tastemaker, editor, or executive I'd invited out for the night. I would host paid nights at hotels or resorts for groups of mommy bloggers in hopes that I could keep them on speed dial should I need to counteract bad reviews for parenting products or manage an online crisis by burying negative press. When I repped liquor companies, I would descend on the city's most popular bars, befriending the bartenders and dropping my corporate card, letting them know my tip would be 25 percent of the bill, plus cash on the side—all from the giant marketing budget instead of expense reimbursement. Bigger tips meant bartenders would recommend us when ordering supplies, serving drinks, and, eventually, when they moved up to management, making their influence more powerful. We'd borrow influence from somebody else's cool event, guerrilla style, by surreptitiously hanging our logos or hiring models to hang out in our branded clothes to make it appear in the publicity photos as if it were our event. I would even crash the best events and parties, trying to connect with the bigwigs I knew would be there.

Every holiday season, I sent customized gifts to each of these people: blankets embroidered with their kids' names, personalized journals emblazoned with their monograms, baskets of treasures from their favorite places and shops. I spent thousands of dollars and hundreds of hours handwriting cards, signaling that I was paying attention, that I took the time to get to know them, that I remembered their kids' names, that I was being thoughtful. I made the gifts feel different from the perfunctory flower arrangements or fruit, cheese, and wine baskets sent by my competition. And that was no accident.

I didn't stop there. To boost my standing, I attended painful boys' club lunches at the Harvard Club and Balthazar's and any other power spots where C-suite decision-makers mingled and toasted with scotch. It was all about making the interactions casual and personal, to promote myself and my clients without ever actually spelling it out—my version of picking up someone else's kids from school and inviting the family to a weekend house upstate only to ask for one *small* favor in return. This is why country clubs have had a long-standing reputation for sparking giant business deals, whether at the ninth hole or over a clubhouse prime rib. In fact, there's now an entire industry built with this "casual" relationship building in mind. Soho House, for example, is a members-only club, built by a committee of key influencers as founding members who invited a list of their key people to be considered for introductions. The Classic Car Club Manhattan is answering the hole in the market for salespeople who can't afford a luxury car or hold the cachet for an invite to the Soho House. For a fee, you can borrow their cars for big business meetings to impress your prospective clients and even attend their events with potential clients. Joining luxury fitness and wellness clubs also increases a person's likelihood of sharing a conversation with a celebrity or powerful person above their pay grade with whom they share the locker room. Conferences and trade shows are transactional events too, where attendance suggests you're ready to talk about future business. And society clubs like the Freemasons, sororities and fraternities, Rotary, American Legion posts, and small business associations bring together like-minded people for the same reasons.

Of course there's a glaring problem here: Not everyone is invited to the table. Many powerful politicians, top executives, and media head honchos were socially engineered to follow paths laid out decades, and even centuries, before, to mix at private country clubs and fraternity gatherings where there was historically a barrier to entry for Jews, Black people, women, and other marginalized groups. And that dynamic is how we arrive where we are today, in a world where the majority of key decision-makers aren't women, people-of-color, or otherwise ethnically, racially, religiously, or socioeconomically diverse. The boardroom is reflective of the clubs and lunches that got them there—the like-minded company they all keep. And when these decision-makers have only ever experienced those rooms filled with people just like them, they make choices based on that tunnel vision—which is great for them but often not for the rest of us. The trickle-down effect of the influence is monumental. This sets structures in place that perpetuate the cycle of elite influence, building upon conscious (and sometimes unconscious) biases and belief systems.

Of course no matter what your identity profile may be, sometimes the influence you wished to wield was just out of reach. And when I didn't have a guy or a girl for that, or a place for a power lunch, or couldn't get by with a little help from my friends, a big idea or big budget was required to break through my tapped-out network and level up into that next sphere of influence. I would stop at nothing to get the sway I needed—just like all those brands influencing you today.

So when tasked with getting one of my brands into the holy grail of elite celebrity events, the Met Gala, I saw it as the ultimate challenge. And I knew my only chance of catching the big fish was to start small. I just needed bait.

There was literally nothing I could do in earnest to participate in such a lofty gathering. The Met Gala is the most talked-about event of the entire year, billed as a fundraiser for the Metropolitan Museum of Art's Costume Institute. But more than anything, it's a runway scene for Hollywood's most famous. And, of course, like all things influential—

especially in culture—it's fully monetized. From the red carpet to the live stream, from the lounges to the photobooths, *Vogue* brings in sponsorships for ad space. The price tag for having a whisper of a prayer of your ad being seen for six seconds alongside the Beyoncés and Lady Gagas of the world starts at $1 million. Factor in celebrities sharing preplanned (and paid for) sponsored branded content, and the price tag gets even larger than the egos. According to the *Business of Fashion,* in 2021, the media value of the Met Gala was $543 million—higher than the Super Bowl. We had nowhere near the budget needed for an ad or a sponsorship. And gate-crashing the star-studded event seemed impossible. So I looked to borrow influence as my ticket inside.

I asked myself: What was this brand really seeking? Was it to be seen in the company of celebrities? Or was it to be mentioned in relation to the glitterati and the aspirational event? What it came down to was the access point the celebrities offered to the media. This brand wasn't star-struck at all. They just wanted to continue growing their brand identity within the luxury clubs so that everyday people (not celebrities) would *mimic* celebrities and choose their brand, seeing it as the most attractive when given a choice.

After some thought, I decided the solution was simple. Where there are A-list stars, there's a school of paparazzi and reporters swimming nearby, hoping to feed their machines—websites, tabloids, gossip shows, etc.—with candid photos, videos, quotes, and interviews. Working outside from a phone or laptop in New York City is loud and annoying. Ambulances squeal, subways grind below, and hordes of people make it impossible to focus. These were my little fish. And I had thought up the ultimate trap.

It was the age of the food truck. So much so, that even mobile clothing retailers had begun fighting chefs and restaurants for curbside parking space. Our brand had not one but two trucks already in our stable. The beverage company was hawking a lesser-known type of alcohol, so we'd hand-painted the trucks, decorated them, and outfitted them with air-conditioning to drive around the five boroughs, offering people samples of cocktails. Wifi wasn't 5G back then. I am not sure it was even 2G. (To offer context, I had a BlackBerry phone.) There were no sidewalk uploads in real time. Access to instant communication wasn't there. The process required phones, computers, wifi routers, cords, and cables. Hell, maybe even a fax machine. *Ding!* Here was the opportunity for a favor. My bit of quid pro quo.

That year, when the stars lined up to walk the red carpet into the Metropolitan Museum of Art on Fifth Avenue while the schools of reporters and paparazzi waited to feed off them, we provided a functioning mobile office down the street. And we offered it to the people we really wanted to influence: the press. We offered a respite from the madness, ease with which to do their jobs better, in the comfort of an air-conditioned—and branded—alcohol truck stocked with cocktails made by our resident model bartenders. We got great press the next day, but, ultimately, that was beside the point in comparison with the lift in brand awareness we'd created with the journalists themselves. That would serve the company for ages to come. We had made some new influential friends.

These transactional relationships, like the one I built between the brand and the press, tended to go both ways. To this day, I still own a blanket that a public relations agency sent me with my newborn son's name needlepointed across it. It was a thank-you for recommending them to a client of mine once, earning them a six-figure contract. They wanted to keep themselves in my good graces should I have another client that might fit. I've maintained convenient friendships with professional photographers, filmmakers, bands, marketers, writers, entrepreneurs, executives, executive assistants (sometimes the *most* important), hoteliers, printers, agency owners, buyers for massive retailers, owners of NYC's best storefront windows, models, and more. If you're in New York long enough, something fun happens. Your B-list friends—bussers, bellboys, and bouncers—become maître d's at famed restaurants, concierges at the most elite hotels, and owners of the next "it" clubs. Your network grows up. Your sphere of influence expands.

The minute you stop maintaining these relationships, though, this begins to work in reverse. As a middle-aged mother of two, very happily out of the scene for the past few pandemic years, many of my "guys" (and girls) have aged up and out too. They've retired, switched professions, or given up the hustle, replaced by new it-girl editors at websites that wield more influence than my magazine editor friends ever did on paper.

The new generation replacing my network may not want bottomless drinks at a club in exchange for a post. Glad-handing and getting to know one another in real life seems to me to have been replaced by ongoing long-term internet relationships, maintained via Twitter conversations and comments, DMs, and LinkedIn communities. They may not like the insidiousness of befriending for benefit. They may appreciate clarity or—*gasp!*—they think they're making actual friendships. Simply put, if you don't grow with the times and continually cultivate your network, your influence shrinks. If you fail to understand the changing world and its impact on the changing minds of those with influence, you may edge toward irrelevancy—and that may be fine with you. Based on where my network was headed, even pre-COVID, I was already a dinosaur.

Transactional relationships happen all the time, and they never look the same. In relationships, it can be about marrying up. It can also look like parental purchases of flashy toys that help socially engineer their kids' early friendships—pools, trampolines, game consoles. For adults, it's boats and summer homes and elaborate dinner parties. In business, it's the corporate card, private club memberships, and opulent gifts. But most often, the common denominator of these friendships and communities is that they were created with the same motivation in mind—in the end, it's about receiving (getting what they want).

The more we become aware of the inner workings of these transactions, of the way people are constantly trading and negotiating with the company they keep, to influence *us* as consumers, the more we can pick and choose who we follow—both online and also more figuratively. Take a moment to stop and consider when being offered freebies or favors; when there's an asset to be loaned or enjoyed; when our elbows are rubbed. When we are in the company of collectors of people, we can pause and ask ourselves: Are we part of the collection? We can begin to understand that we're being spoon-fed the trends, brands, and tastes we embrace because of backdoor dealings, not because of some inner compass. We can open our eyes to the fact that we may like or be aware of something because someone three degrees from us was targeted and is unwittingly spreading the word.

Not all connections are made with ulterior motives. Sometimes a friendship develops naturally out of a rapport, respect, and commonality. I've befriended some amazing humans in my lifetime, who I have wholeheartedly lifted up whenever I could—not out of obligation but because I believed in what they were doing, making, saying, selling, writing, and creating. Therein lies the difference for me these days, I think. I'm no longer creating networks for networking's sake. There's no more tit-for-tat mindsets. I've found the important difference it makes to surround myself with a network that's positively built. I'm befriending people whose voices and ideas I believe the world needs. I hope those friends feel the same about me, and I hope you will choose to surround yourself with people you believe in—and who believe in you too.

Even for a cynical manipulator like me, these relationships, born of true mutual respect, are the best ones and those I have chosen to keep in my dinosaur years.

Will I ask them for some borrowed influence as I release this book into the world? Ask them to have me on their podcast or write up a blurb? Will I find connections to my inspirations—writers, thinkers, and unabashed freethinkers—who are less than six handshakes or one intro email away?

Hell yes, I will. After all, I've got big fish to fry.

Folklore   It's all in who you know.

Folktale   I don't know the right people, so I won't get what I want.

Folk Rebellion   By looking at people I truly like and admire, I can grow my circle strategically and transparently but also authentically, so that it makes these relationships long-lasting and built out of mutual respect and friendship.

*Raise Hell   Draw a map of the people you know. How did they come into your sphere? Did they arrive there by happenstance or circumstance, or were you (or they) hoping to gain something by becoming contacts? Of these people, who do you hold close and dear to your heart? Which ones remain on your "phone-a-friend" list when the world falls apart? Now see the ones who are left. My guess is that they are the more tit-for-tat acquaintances. Noting what role each of the "others" may fill in your life, what would happen if you tried to find more "near and dear" in the tit-for-tat relationships?

ROSY FUTURE

(OR Nurturing Interests
Instead of
Feeding Trends)

I've got a weird little gift. I'm able to see and predict things before they happen.

This is a mixed blessing. On one hand, it means dancing my cares away to the music of the times. On the other hand, it means ignoring the nagging fear of our impending existential dystopian doom based on that music. It's dancing on the outside, dread on the inside.

For better or worse, I can read minds.

Not in the metaphysical sense, but in the read-between-the-lines observational way. I notice patterns, repeated events, or things gone missing that were once there. It's why I could sense the resurgence of disco vibes even before Beyoncé dropped a futuristic disco album. In a world that had become much less upbeat, I could see the need for a kind of "fuck it fun." Disco was the only clear answer to our pandemic despair.

The same way I became adept at reading people in person, when life moved online, I began to realize I could read people's cues through the screen as well. My ability to find meaning in the negative space translated to what was, and wasn't, being said or shared. Instead of interpreting body language and verbal cues to discern what people were thinking or feeling, I found myself clocking what they did or didn't post. I considered the meaning behind lifestyle cues and changes in behavior like if they were working out more or working more—and began to assess what that likely meant was happening in their lives. The more I've leaned in to this, the more I've realized how much can be discerned from posts intended to project something benign or unrelated.

About to be dumped. Looking for a new job. Moving away. Going to become a life coach.

To be clear, I don't like knowing these things. I'm not seeking the info; it's just how I've been trained. It's like people speaking in another language for privacy, in which I am fluent. I don't want to overhear and understand, but I do. It's also a little like wearing the opposite of rose-colored glasses, and it makes me wonder how much people understand about what's about to happen to them. If certain patterns—overcompensations, style shifts, changes in routine, etc.—are obvious to scrollers like me, does that mean the people posting also suspect they're forecasting their reality behind the facade? Do they know a breakup is imminent too?

I'm not so sure. Being perceptive in this way doesn't seem to work inwardly for me. I've been walloped to my knees with huge blind spots. People are inclined to make excuses for when we miss glaring realities about our own lives. For me, they said, "You're too close to it" when my boyfriend was fucking the neighbor, unbeknownst to me but apparently clear to others. They said I was being optimistic when I collaborated creatively with someone who, in retrospect, clearly resented me. For a lot of the time, I think we only see what we want to see—and what we can handle facing. But from a distance, for others, I'm almost never wrong. Sometimes it downright blows.

But there are other ways in which it can be a superpower—especially in business. Being alert to micro- and macroshifts, in the larger culture and in people's individual lives, makes me uniquely qualified to gauge how people will react, which means I can often predict trends (disco, people!). My intuition rarely fails me. Of course it wasn't always this way.

It all started with a big mistake. For the sake of generosity, let's call it "a teachable moment." My lower back tattoo. Yup. I have one, a *time* stamp of when I came of age—in the early aughts. I had what I thought was a very original idea: a tattoo above my butt crack. No, I am not joking. It seemed unique to me at the time. But looking back, I now realize that I was just influenced by the low-rise denim fashions of the noughties. No sooner had I got one than I started seeing them *everywhere.* Y2K icons Pamela Anderson, Britney Spears, and Christina Aguilera were all getting *tramp stamps,* a term coined by a *Saturday Night Live* skit and forever linking my ink to its negative connotations. Had I sensed that this aesthetic choice would become a parody of its own time, I probably wouldn't have gotten that Chinese lettering (*groan*) all across my lower back.

My Spidey sense was off in those days because understanding how "now" came to be, as a product of the influences of the past, is like having an already baked cake and trying to determine the ingredients inside. In my case, the recipe was hip-hugging denim and crop tops, pop culture feminism, and MTV. Defining a decade, or a zeitgeist—a mood of the times—is something that most often happens in hindsight, after looking back at the trends, stories, and culture of the era years later. Because until you can see all the ingredients, you can't know if you even have a cake.

Anticipating the future, on the other hand, is a different beast—and, ironically, it's simpler than understanding the present while you're in it. It involves examining what is happening now and presuming how that will influence what's next. It's having the ingredients and sketching out how the cake will look. And like baking, there's a recipe and there's guesswork, all open to interpretation, instinct, and unpredictable outside forces.

People get paid to do this—predict the future of tastes and trends. And after my harrowing tramp/time stamp incident, I became one of them. Having the ability to sense things coming, based on the ingredients I observed at hand, allowed me to build not only the cake—but a career.

And, ultimately, as a brand strategist, I would get paid to both sense trends *and* push them. Because while interests may arise organically, fads that catch on most often also involve outside forces intentionally nudging the culture in a certain direction.

When I saw aging executives at the highest level in their careers becoming nervous in meetings about their inexperience with the internet, I parlayed that insight into my Viral Networking class. That, in turn, led me to become a social media and digital strategist. I became this sort of liaison between youth culture and the C-suite, which didn't understand this new digital world whatsoever.

I came of age during a very specific moment of change, allowing me to remember life before and after the internet, thus understanding how both traditional marketing and the digital world could converge. I was uniquely part of both cultures. So I sat at that nexus, serving up information about the modern world in language that traditional executives could understand. The world was hurtling toward its digital destiny and decision-making dinosaurs would throw money at anything if it helped them feel safe in their jobs. Ultimately, because of my unofficial training in the art of influence and manipulation, and because of the moment at which I joined the workforce, I fell face-first into a career that hadn't existed before. I saw the trend happening, and I used that trend to push other microtrends that served me.

Futurology, trend-spotting, and forecasting are real career paths, and they involve some self-fulfilling prophesizing. Seeing a trend emerging and then giving that information to companies who then capitalize or double down on it in order to appear relevant is actually cementing that same newfound trend. That's how waves of fanaticism go from *subtly* to *suddenly*.

Popular tastes are often reactions to pendulum swings too. Today's trend toward noncolors and sad neutral everything—toys, homes, fashion—is a reaction to the prior decade's millennial pink. This *suddenly* phenomenon is why one day you look up and it feels as if overnight everyone is planking, drinking White Claws, or posting photos of orange Aperol spritz cocktails. It's why, suddenly, every restaurant in your neighborhood has ramps on the menu.

Trends may seem as if they appear from nowhere, but they don't. And the ones that spread like wildfire instead of simmering and burning out were 100 percent selected and pushed by companies, intentionally. I can sense what's coming because I've been trained to be attuned to people's desires and inner motivations from all my previous forays into behavior modification. But it's when people like me give that information to and share that instinct with private companies, that a growing interest turns into a giant megatrend.

At one point, you may have thought it was a coincidence that you and your best friend were suddenly both obsessed with drinking sidecars, but actually old-school craft cocktails began trending as an answer to mass-produced beverages. The popularity grew out of a slew of old-timey bars and speakeasies that popped up, leaning in to old nightlife classics like dark lighting and limited seating and a bartender who knows your name. This was in opposition to the superficiality of "drink and drown" frat-style, fast casual dive bars (like the ones I used to work at!). But the interest in a more upscale, old-fashioned cocktail experience actually signals a larger trend at work—a return to simpler bygone days. (You can also blame this trend for the proliferation of hipster handlebar mustaches and pompadours, fades, and side-part undercuts.) And that's *all* because, deep down, we started to feel overwhelmed by the crazy stimuli of the world with everything going online, becoming nonstop and connected, and being superficially *extra*. We needed to slow down. (Our desire for neutrals these days may also reflect that, to an extent, there is a sense of searching for calm wherever we can find it.)

In opposition to such overwhelming stimuli with the advent of the internet, we entered the age of everything "authentic," "craft," "bespoke," and "handmade." This spawned phenomena like Etsy, which began in a Brooklyn apartment in 2005 as an online virtual storefront—an "artisanal marketplace" for arts and crafts. It became tech's poster child, the largest venture-capital-backed IPO in New York City's history. It was originally conceived as an online platform for makers to sell their goods, which is still what it is today. Following in the grand tradition of craft fairs, Etsy gives sellers personal virtual storefronts, where they hawk their goods for a fee. This was a perfect example of an idea catching the wave of two larger cultural trends at once, which might on the surface seem diametrically opposed: the rise of the internet and a return to basics. And it grew massively. As of the writing of this book, Etsy had 120 million items in its marketplace, 7.5 million sellers with 96.3 million buyers, and 2,402 employees. In 2021, Etsy had total sales of $13.5 billion on the platform.

The *now* megatrend—all things artisanal, handmade, and of simple quality—is big business. That once small craft marketplace now challenges the more obvious capitalism of ecommerce brands like, say,

Walmart, ranked only a few spots above Etsy in terms of traffic. In fact, Etsy blows walmart.com out of the water in terms of returning customers. So as dinosaurs do, Walmart is now attempting to jump into this slow-shopping, simpler-way-of-living movement a tad too late. Deeply off-brand, the huge chain store has now partnered with a smaller escape and retreat brand, garnering countless disappointed comments on the "authentic" brand's social media. The partnership is intended to encourage consumers to power down, disconnect, re-charge, and "Save Money, Live Better," as Walmart's slogan says, in the great outdoors. The customers' desires shifted over time and *trended* more toward authenticity, connection, and simplicity. So what's a megacorporation, the poster child of late-stage capitalism, to do? Borrow (buy) influence, of course.

Trends often start at the root of the zeitgeist, in something that is happening all around us, and grow from there. And eventually, de-cades later, the megatrends that were born out of that pain point are co-opted and capitalized on by the megacorporations. Then, just like that, the trend becomes passé.

Tracking trends, creating them, or participating in them doesn't necessarily help us, even if they're rooted in trying to solve a pervasive issue. The problem is that the trend's emergence doesn't always solve whatever pain point inspired it to emerge. And engaging with or pro-moting that particular trend so that it grows in popularity can push us further from solving real, society-wide issues that affect all of us who need real, dedicated solutions. Someone developing a craft cocktail or online space for craftspeople isn't necessarily benevolently thinking the world needs to slow down or that people need to feel connected to real things made by real people in order to unplug from all the stimuli. But becoming aware of how these trends emerge and who benefits from lighting a fire under them just might help free us.

In my life, I have benefited greatly from being able to stay ahead of the pack. Each line item on my résumé demonstrates just how much I leaned in to this ability to spot what's coming and then sell those real-izations to my clients. It also shows how I fell prey to the zeitgeist trends myself.

My being an *early adopter* of digital got me work as a strategist for *emerging brands,* which put me at the *forefront* of making lifestyle communities for my *innovative* clients, developing them into category *leaders*. And by pushing the trends, I became a *visionary* for our over-stimulated world and a *pioneer* in the slowdown movement, *trailblazing* the need for digital wellness as a *female founder* in the betterment zeitgeist. WOOF. I was like one big buzzword. What a load of crap I helped create!

My back tattoo was a hard lesson learned (and a permanent reminder)—trends always pass.

Brands (and their marketing teams) can be blamed for pushing things into the cheugy phase, when they become the opposite of trendy. When we, as a culture, are ready to move on, we see once hip concepts featured in too-slick pop-up ads, then for sale on the shelves of every discount store. I fear my beloved Brooklyn may have already jumped the shark. Once it was a bastion of crafters, craft pickles shops, craft cocktail bars, and craft breweries. Now I overhear the bro who just moved in across the street saying things like, "Saturday is for the boys!" Am I being pejorative about the lifestyle trends of people younger than me? Yes. Yes, I am. Like a true elder, my disdain for how hard they try to look like they're not trying is intense. Literally, there goes the neighborhood with their normcore nonfashion, and Warby Parker–style dental offices. Just as I can spot a trend, I can also see one passing. Montauk, the Catskills, and now Brooklyn. *Dead*. Or at least my beloved version of those places.

But sometimes that's a good thing—having fads truly disappear. At their worst, trends become toxic structures in our lives. Sometimes, as we've explored earlier, pain points are invented in an advertising agency or a boardroom to sell a product as a cure. That means we're convinced we have a problem that we never thought we had! Or if a societal trend is pure and began as an important movement, it will inevitably be gobbled up, bastardized, and commodified for profit.

I have personally witnessed the falseness and hypocrisy of these dreaded lifestyle trends and brands from the inside out, as have many of my friends. Ask most women working for brands outwardly championing the work-life balance for moms, and they'll tell you the same companies are edging out the employed working moms from the inside. Organizations who build their businesses by lifting up creatives often steal from those they claim to support. Wellness media conglomerates that tout every new self-care offering—mindfulness, "health is wealth" mottoes, juice cleanses, Pilates sessions, organic mushroom teas and supplements—have incredibly high employee turnover rates due to burnout and cortisol-induced autoimmune diseases. Most absurd is the slowdown start-up advocating for a less-is-more lifestyle, while forcing employees (or in my case, me) to burn the candle at both ends behind the scenes.

I've been sucked into—and touted—enough of these brands to know that even the well-intentioned movements eventually get corrupted and then evaporate into the ethos. Every single one. The tech era cool "small" lifestyle brands, the hustle culture, chief mom officers, *she*-EO, *Lean In,* simple living, Girlboss, millennial marketing, and self-care-first all have or will peter out, leaving behind only the mug that says Rise and Grind. They'll be forever immortalized in their space in time, just like my Chinese character tramp stamp tattoo, which not so ironically (supposedly) translates to: "Live for today." You can't unbake a cake.

When the pendulum inevitably swings the opposite way, we are often left with regrets—whether bad haircuts or lasered full-Brazilian pubic hair that won't grow back with the next bushier trend. It's not bad to follow trends per se, but it's not the same as developing actual tastes or supporting an actual movement. Wearing millennial pink is one thing; building a pink palace in every city as a hub for women's empowerment through community is another. Sure, join the pink club if you want the connection of fellow ladies hustling at the grind of entrepreneurship. *Lean in* if you enjoy operating in the system originally made by men and want to conquer it. Work for the tech monsters if you plan to cash in and sell out. Proudly post your passive-aggressive "gentle parenting" lessons on your TikTok page as you wax on about the joys of simple living while wearing a *Little House on the Prairie* "nap dress" and watching your kids play with their beige toys if it helps you "spark joy." Play Wordle—not because everyone else is doing it or you want to show how smart you are—but because it's fun.

Wouldn't it be better if we were free to develop affection and interest in the things we were naturally drawn to instead of what we're being unconsciously spoon-fed? It's important to grow in our awareness that sometimes an interest is based on a trend. It's not necessarily bad to gain a new interest that's ignited by a trend, but knitting having a moment because people so desperately needed their hands to do something besides scroll is a perfect example of a trend pendulum: so many who partook in the hobby found that, once it faded, the feelings and fears they were avoiding by this trend distraction still remained. Sometimes your "unique" tramp stamp is just the result of unconscious influences. To truly develop your unique self, fostering a mindset of choosing things because they appeal to you, not because it's what everyone is doing or because of a problem the trend claims to solve, is key.

The route to thinking for yourself begins with recognizing the influence of what's so hot right now and what's really being sold, and making conscious decisions based on that rosy—not millennial pink—awareness. And before it gets too popular, it begins with a little disco dancing too.

Folklore   Accepting trends is a normal part of being human.

Folktale   This is all the rage, so I have to join in to fit.

Folk Rebellion   I don't need to follow the trends unless I have decided they are good for me, I like them, and I choose intentionally to adopt them.

*Raise Hell   Your haircut, your clothes, your music. When you look back on them, are you an early adopter or a laggard? Are you buying full-legged pants because skinny jeans have become passé, or will they bury you in your Uggs regardless of the decade you die in? Do you slowly or quickly follow the masses? Or are the things you wear, do, and think based on your own likes? Can you name three interests you partake in now because you enjoy them, and not because of a trend that influenced their popularity? Are there any new interests you've been itching to try that are outside of any existing trends? Can you give them a try now?

UNPLUGGED & LOVED

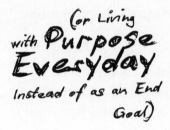

with Purpose (or Living
Everyday
Instead of as an End
Goal)

When you are a truly bad influence, you can be so bad, you trick yourself.

It's taken me almost a decade to realize that *Folk Rebellion*—the brand I created theoretically to help others unplug and connect to what's real—was born, thrived, and died because of my unhappy marriage. I was grasping for a sense of fulfillment but also an escape from my everyday life. If I could stay busy enough, helping others find happiness, then maybe I could ignore what was really going on in my own life.

Coming to terms with the truth crystallized how I would live my life going forward.

I will not take away from the fact that I truly was "driven by a mission to raise hell for living well in real life (#IRL), encouraging people to become mindful of the technology they use," as stated in the company's press release. But understanding how influence works—through my lens now—has given me the hindsight to see my true motivations clearly. Looking back from my standpoint today, and realizing I had an enormous need for something to pour myself into as a way of avoiding my reality, I know I wasn't championing a cause purely for honorable reasons no matter how much I truly believed in it. That reality check has been both pivotal and sobering.

At the time I created *Folk Rebellion,* I was unwell. And that unwellness laid the groundwork for my creative baby, *Folk Rebellion.* I had convinced myself that my physical and mental fracturing was due to the negative impact of tech—and I stand by my beliefs about the dangers of a nonstop digital dystopia. But I see now that it wasn't entirely tech's fault, as I had believed, raged openly about, and built a business upon.

Most of us now see that in the sixteen years since Steve Jobs put the world in our pockets, technology has crept in and commandeered everything from our attention, to our connection, to society. And maybe that hasn't been the healthiest influence! But it's taken us this long to even start to realize that.

As an early adopter in the digital realm, at the forefront of sitting in front of screens for money, I was like a guinea pig, and I experienced what would today be considered accepted repercussions of too much screen time. In 2012, I began developing ailments that were essentially pooh-poohed as female hysteria. When I said that I was feeling truly bad, no one took me seriously. Doctors, a holistic practitioner, a boss, old friends—all at a loss about my symptoms—suggested the causes might be poor diet, lack of exercise, postpartum depression, and autoimmune diseases, none of which explained why I was confused, disoriented, detached, struggling with word retrieval, lethargic, foggy, and forgetful. They seemed to think I was just being emotional—as if I couldn't tell the difference between feeling a little sluggish after eating french fries for lunch and experiencing full-on vertigo.

Admitting that you feel as if you're losing your mind instantly puts you at a disadvantage. Because, if you are right, then why should they believe anything you say? I was in my early thirties, otherwise healthy, and my lab tests all came back inconclusive. So nobody believed that something was wrong with my brain (except perhaps my emotional grip on reality). In other words, they were sure it was all in my head.

It wasn't until I was well into my scrambled-eggs-for-brains phase that a neuroscientist named Manfred Spitzer coined the term *digital dementia.* Unfortunately, I didn't actually read about this notion until 2015, when the media started to call attention to it. Spitzer described how an overuse of digital technology could result in the breakdown of cognitive abilities. He likened this phenomenon to symptoms more commonly seen in the elderly or those who have suffered traumatic head injuries and/or psychiatric illness.

Thanks in part to the informal intervention that my onetime husband sprung on me on our family trip to Hawaii, I didn't have to wait to read Spitzer's theory before recognizing one of the sources of my troubles. I'd been lucky enough to have my veil of unwellness lifted long enough for me to pinpoint its source. Root causes tend to show themselves if the very thing causing the ailments (like tech, alcohol, busyness, capitalism, relationships) is abstained from, even if for a little while. I could trust what my body was telling me. I didn't wait for researchers to catch up. I simply opted out of what was making me sick and started anew. But once I read about digital dementia, I felt vindicated. I had the scientific proof to back up my personal experience!

My previous career was very fast-paced and kept me almost wholly online. As an award-winning brand strategist and creative director for rapidly growing companies, I built digital ecosystems and community through storytelling and communications. My background in the online space gave me a perspective not many on the inside were willing to voice at the time.

I am still, as I like to say, a recovering strategist. It took many years and a million tweets, status updates, emails, and living underneath a waterfall of data for my brain to finally give out. When I had my great awakening about the influence the internet, digital and hustle culture, and corporate America had on me, I was able to muster the courage to unplug from it all, almost instantly. (Sometimes my more impulsive side works in my favor.) I realized that not only was the time online making me sick, but I was also living under the thumb of a culture that told me that incessant screen time was the secret to success and possibly even fulfillment.

My newfound freedom from that influence blossomed into a wildly creative streak. Always one to go big, I made it my mission to educate the world about the perils of technology and innovation without boundaries or balance. If my feelings of listlessness, fatigue, memory loss, malaise, and brain fog were all related to my overdependence and possible addiction to technology, I shuddered to imagine what the world might be like for my son as he got older if we stayed on this current trajectory. Looking back, I had a whole offline life before my twenties, filled with hobbies, friends, time in nature, and real analog experiences—and yet I still fell prey to the powerful influences of the tech world. What would happen to this kid, *my* kid, growing up without ever having known life without these influences? How would he and his generation cope?

So at first, I did what I knew how to do best: I created a lifestyle brand in the hopes that I could positively influence a world of consumers into waking up to the realities we were facing. Maybe then other people could join me in enacting change. During my whole professional life, I had built communities around brands, so I was never really selling a product but rather a lifestyle. I mean, I knew no one wanted to be told they were scrambling their brains and, worse, their kids' brains too. I had to handle this smartly, maybe even surreptitiously.

I knew that if I shared research, it would only be read by people interested in research. If I reported tech, my content would mostly be read by people in tech. And academic or futurological books and reports also live in their own echo chambers. If I wanted to reach the everyday person (or *user,* as they're known digitally), I needed to be where they were and speak their language. It might seem counterintuitive to get on social media to take it down, but you need to fish where there's water. So tapping into my marketing brain, I decided the best way to get people's attention was through a *feeling,* and the best place to get in front of them was on their newsfeeds.

People's relationships with their technology are very personal, as was my message, so I needed to make the content fun, lighthearted, and approachable—never judgmental. I needed a spoonful of sugar to help the medicine go down. So I conceived *Folk Rebellion* as a countercultural, rebellious, and edgy lifestyle brand to spark a conversation, hoping to wake people up to why they were feeling off. I would talk less about the technology and more about what's lost or gained from our relationship with it.

In conversations in my personal and professional life, I kept hearing a recurring theme: People were sick of their loved ones choosing the screen over their company. They were burned-out from the incessant dings and pings in their own lives too, but were still more distracted, overwhelmed, and addicted than ever. At the time, either I underestimated how many people felt like I did, or we didn't have the terms for it. You have to realize that *digital detox* was barely in the lexicon back then and yet more and more of us were feeling overwhelmed by the tsunami of the digital revolution as we began to understand its downside. In the early days, when people asked, "What is *Folk Rebellion?*" I found it kind of hard to explain that I was trying to remind people what it was like to be human again. But they dug it.

Folk Rebellion's mission was "to be the alternative for a screened-in world, leading the plugged-in toward digital well-being." Back then, I believed society *could* live well with technology through a more informed, balanced, mindful, and questioning approach. (Today I'm not so sure—the system is rigged against us, so I won't play by its rules.) Through that lens, I worked to explore the crossroads of humanity and the new modern world, and through editorial content, experiences, and partnerships, I attempted to monetize the brand so I could pay the bills. I quickly amassed a community of more than 150,000 "rebels." My ideas became one-liners on T-shirts—UNPLUGGED AND LOVED, WANDER WHERE THE WIFI IS WEAK, TAKE IT EASY AND OFFLINE, SO MANY LIKES WHEN ALL YOU NEED IS LOVE, HOLD SOMEBODY LIKE YOU HOLD YOUR PHONE. The tags were bookmarks and friendship-bracelet-making kits with messaging for pushing back against technology, offering an analog alternative to their phones. They were sold in aligned retailers like surf, coffee, and yoga shops, plus well-known national stores. I became a regular fixture on the podcast circuit, the speaker circuit, and the panel circuit. With *Folk Rebellion,* I used my experience and knowledge from the career that had burned me out to spread the good word. I'd hear from nurses who started "unplugged clubs" in offices to reconnect with their patients; I'd get snail mail with handwritten thank-you notes from people of all ages; and on my Off-the-Grid retreats, I'd hear personal stories of how tech was affecting people's kids, work, health, and relationships. I was humbled by how quickly *Folk Rebellion* took on a life of its own.

The catalyst for starting *Folk Rebellion* was prioritizing my own mental health, but my impulse to run away from myself, my emotions, and my reality—by diving headfirst into the project—helped it flourish. Ultimately, I wasn't actually taking my own advice. I was killing myself while preaching to others. Instead of sneering at the wreckage of my marriage, I was sneering at the wreckage of our society—and that was somehow easier.

It was a welcome distraction.

Many creations are born this way. There's been a long history of artists, writers, musicians, entrepreneurs, founders, and inventors finding their voices in periods of desperation, sadness, loneliness, and poverty. Still, it goes to show you: Even when your intentions are improved, your influence may not be what it appears to be from the outside.

I'm no Van Gogh. My hurt wasn't channeled into a great art or even a bestselling album. But I found it easier to turn to my computer, still feeding my tech addiction, and lose myself in the creative process. It was easier to work on something I intrinsically loved rather than face the problems at hand. That's how many of us wind up losing years online—it's easier to scroll through other people's lives and stories than it is to take a hard look at our own.

I loved thinking about the new digital world. I found joy in making something that never existed before. To me, problem-solving is fun and challenging. And, like Etsy's founders, I felt like there was space to both embrace an old-school ethos and invent something that thrived online. These were my intrinsic motivations—or what was personally rewarding to me and not about some external reward—when I first started *Folk Rebellion*. My intrinsic motivations were, and still are, pure and true.

Unfortunately, they didn't remain my only motivations. I was also influenced by external factors outside of my own sense of self-determination. Our motivation is rooted in our extrinsic reward, whether it's a paycheck, good grades, or trophies. The external intangibles are sometimes even more tantalizing—rewards like praise, recognition, or attention can steer our lives in the wrong direction. Reaching external summits—*when I buy that car, once I'm promoted, after I receive that award*—brings only a short-lived sense of satisfaction. That fulfillment is outside of yourself and, therefore, temporary. It's devoid of the intrinsic fulfillment of just doing it *because*.

We all know people who are forever striving. Never at rest. The things amassed never enough. The carrot keeps moving, yet they keep chasing—and the goals keep growing or the goalposts keep moving. These individuals are never happy because they are attempting to fill the void externally.

I am not usually one of those people, at least not at the core. I have been able to fill my own cup for as long as I can remember—since back in the day when I bucked my community's expectations and went to work at a ski lodge instead of going to college. But my background will always influence my present, and in the times when I have gotten offtrack, I've been able to recognize the issue and course correct. But something interesting happened with *Folk Rebellion*. What was joyful for me was also born out of my need to avoid pain. Both could be considered intrinsic. But the more I did work that filled me intrinsically, the more I was externally rewarded, leaving me, ironically, with less joy from the process.

Soon I was sitting in meetings with buyers discussing color trends coming down the pipe for our T-shirts, just like I had with my branding clients. I hate fashion and I hate trends. I began meeting with megabrands, creating presentation decks filled with my ideas for pop-ups, events, podcasts, TV shows, panels, trips, and corporate wellness programs. Every engagement was transactional. Could I be their spokesperson if they started a digital wellness offering (which they obviously should if they wanted to stay relevant)? Would I consider consulting as they reimagined their brand? What if my concept for an IRL podcast series was sponsored by them, but they owned it after the first season—would they fund it then? If I was to keynote a conference, would I introduce them to my media contacts?

These corporations, founders, and businesses dangled money, notoriety, growth, scalability, and contacts to get me into their offices. They provided lots of external motivations to influence me, and then exploited me. And I fell for it. Hook, line, and sinker. First, my newspaper was ripped off by a global giant that offered to support me but stole from me instead. For nine months, I assisted a marketing tycoon at a giant wellness brand, only to be told that the tycoon had left, and my work would live on without me and without compensation. I pulled together Skype calls with a smattering of different aligned partners who pulled out at the last minute, only to later organize exactly what I had pitched, and without me. I was supposedly escaping toxic tech, and yet I was still swimming with the worst of the sharks.

That last and final year of *Folk Rebellion*, as it sat poised to become the next big thing, I was bled dry. I was creatively, financially, and emotionally in debt. And this time, I had only myself to blame. I'd invested in external carrots—spending my time doing unfulfilling things like discussing fucking silver lamé T-shirt text—instead of doing what lit me up—writing, advocating, creating. I'd fallen into my own trap.

In truth, I often feel trapped by my projects once I put them into the world because of the tidal wave of responsibilities that come with them. It's a not so uncommon experience for those in the business of "finding something you love to do and never work a day in your life." Apparently, there's a paradox when it comes to getting paid to do what you love. For instance, if a person loved making art in their spare time (intrinsic motivation), and then they got hired to regularly talk about working as an artist and started receiving speaking fees (extrinsic factor), a phenomenon can arise called the *overjustification effect:* The artist's intrinsic motivation could diminish. In other words, the artist can begin to see the speaker fee as more significant than the creative process. I am not saying this is true for everyone, but it proved true for me. Something I really enjoyed was rewarded so often with things I didn't find important, fulfilling, or beneficial that I fell into the trap and lost interest.

It's at those moments when we are feeling pent-up, overworked, and unfulfilled that we should pause and ask: Are the actions I'm taking—the thing I'm working toward—internally or externally motivated? Consider one of the goals you are striving toward right now, and compare it against the two categories below to see:

Intrinsic Motivations
- [x action] makes me feel happy, fulfilled, energetic
- [x goal/action] matches up with my values
- [x goal/action] helps me feel connected to myself

Extrinsic Motivations
- I am doing this because someone asked me to.
- I am doing this to pay the bills.
- I am not finding joy in my actions.

If you checked your current actions (like, maybe, reading this book) against this list and you got "intrinsic" as your answer—Great! Keep going—you can do it! If external, as often it is, it's probably time for a pivot—and we'll cover what that looks like in the next section.

Eventually, after my health began to suffer again and I collapsed at Logan Airport, I had no choice but to face the truth—about my intentions, about my state of mind, about my marriage. While my relationship might have once worked, it was no longer serving either of us. It was time to make a clean break from many elements of my old life— and that included shutting down *Folk Rebellion.*

When I looked hard—and honestly—at the influences driving my motivations, I could see a mixed bag of iffy intentions. All the achievement in the world didn't matter because the hole I was trying to fill still remained. I was still trying to succeed at all costs, ignoring my own values. When the creativity was replaced with accolades and the joy was diminished by to-do lists, I decided to step back and practice what I was preaching. What remained was myself, my unhappiness, and a newfound perspective. I had to walk the walk and talk the talk. I thought I was on the path, but really, I was only nearing it—still caught up in the outside influences in my life and someone else's metric for success. A great and easy way to distinguish between intrinsic and extrinsic motivations is to ask yourself who is defining your success. (In life or in marriage.)

With *Folk Rebellion,* I really truly believed I could make a difference. I thought I was using my powers for good—and, sure, that I could make a buck while doing it. Fundamentally though, I was re-creating the same toxicity I had been living with for the previous decade—the internet as a megaphone, busyness, status, ego, influencing, lifestyle branding, consumerism, and more. I was such a bad influence that I had influenced myself. But this time it wasn't on behalf of someone else; it was me—my idea, my name, my brand, my business. This is what can happen when we play into the bad influence: We end up playing, and hurting, not only others, but ourselves.

When I closed it down, I wrote one final post in February 2020:

"Okay. I think that does it!" I wrote. "Life goes round. We succeed. We fail. We hang on by a rung. We open, close, and shed. And sometimes we arrive right back where we started from. Twenty years ago I moved to Brooklyn, quite literally around the corner from where I now live, to become a starving artist. Better late than never."

It wasn't until I let it all go that I could do what I was most afraid of doing—the thing I'd wanted to do since I was a kid—write this book.

Folklore   Having a purpose is the key to life.

Folktale   I need to find my purpose and live it fully.

Folk Rebellion   Finding your purpose as an achievable goal is a myth; living life purposefully every day will lead to greater happiness and well-being.

*Raise Hell   Write down some of your goals. Those you've met, those you've failed, and those you're working on. Were they extrinsically or intrinsically motivated? For the extrinsic ones of your past, can you see if there's a long line of never-ending met goals after it? If so, what is your endgame? Is there an end? Or is there always a new carrot? Ask yourself why. Sometimes the call is coming from inside the house. What is one intrinsically motivated goal you could pursue in the future?

To be nobody-but-yourself—in a world which is doing its best, night and day, to make you everybody but yourself—means to fight the hardest battle which any human being can fight, and never stop fighting.

—*E. E. Cummings*

Part IV

Not So Easily Influenced

Now that you have the knowledge of *how* influence works, *where* it lurks, and *why* you should be careful with your own motivations to wield it, you may be waking up to a whole new reality. It can be very exciting to decide that, yes, I am going to live above the influence! You'll see possibilities where you were once programmed. You'll begin rethinking the world around you. Your new inner rebel will start stripping away the confines of convention, letting go of old ideas and making room for your own.

The idea that "it is what it is" is no longer something you will embody.

This idiom has always been a way for people to shoulder-shrug their way through life. They're throwing their hands up and saying they accept the situation as it stands. That the outcome cannot be changed. That the status quo is good enough.

"It is what it is" has become the collective mentality of our culture and one I believe is in large part aiding all of our societal problems. It allows us to push off responsibility. It allows us to accept something less than stellar, no matter who is losing out. It stops the thinking process dead in its tracks. There is no room for solving issues or creative thinking about how something came to be or how it can be changed. It makes wrongs seem inevitable or losses unavoidable. "It is what it is" says that a problem is so hard that you're best not even trying, that the influences layered upon one another are so powerful that we must simply succumb.

But that's not you anymore.

You now know that accepting things as they are, never questioning, is exactly how we ended up here—tuned out, passive, uncaring, unaware, unhappy, unwell, consuming, and never changing—and you're no longer okay with that.

But what happens when your first forays into defying influence and living intentionally are challenged? Because—guaranteed—they will be.

To step outside of our structures, systems, norms, rules, beliefs, and accepted way of behaving makes people doubt themselves for remaining within it. It also stops the cycle of influence where someone who benefits from your thinking a certain way, your following certain rules, your disliking yourself, your believing there is always more to do or more to buy is no longer able to benefit from your compliance.

So yeah, you bet your fucking ass there's gonna be some feathers ruffled. (That's usually when you know you are headed in the right direction!)

Balancing our newfound intentional choices against the pushback we may encounter is necessary. In this section, you'll hear about the hurdles I faced most often in my journey and the common red flags you can watch out for too by walking through the prisms of pushback of my own experiences, missteps, and realizations. You'll also see how I recognized moments in which I needed to reevaluate my choices and make adjustments as I found my best path forward—and you'll learn how practicing the skill of reassessing and adjusting as needed is important and valid too.

The following essays explore how I grappled with resistance when I've made commitments to live intentionally—above the sway of influence—and how you can practice staying strong and resolute when rebuffed, discouraged, and rebuked in your intentional living too!

You have all arrived here from separate starting points and come at it from varying directions, but your journeys have now converged here. So come on! Grab your tinfoil hat, your renegade flag, hold your fist high, and shout with me:

It is what it is—till it is not!

HANGING UP

*(on using **Technology** as a **Tool** vs. Being Consumed by Tech as a Lifeline)*

If you were to open the cupboard next to the fridge in my childhood home, you'd find my mom's collection of handwritten pledges declared by her children, the kind of pledges a mother knows will not stick. Something we siblings have come to call the Wall of Shame.

It's been an inside joke for decades. If my mom heard one of her kids make some kind of audacious declaration, she had us write it down on a slip of paper, date it, and sign it. Then she'd tape the pledge up in the growing archive of the museum of naivety and delusional optimism. If we tripped up, we had to return to the cupboard and record our offenses.

Some of my more benign ones fit for print:

> "I am never drinking again." June 11th, 2006 (1st offense: June 12th, 2006)

> "Jess will not set foot in Props Inn (our local pub) for December 2008 or will pay Mom $100" (1st offense: December 3rd, paid. 2nd offense: Dec 6th, paid *Moratorium on future payments)

> "This time I'm staying. I'm not moving again." (Offenses: 2001, 2003, 2005, 2006, 2009, 2012, 2017)

Revisiting the Wall of Shame is a cringeworthy, hilarious, and sometimes painful experience. Next time I visit home, I will add the following to the exhibit:

"I am never using my iPhone again" Pledged December 2014 (Offenses: 2015, 2018, 2021)

Like every other pledge, this latest one was born out of the righteous indignation of youth: certainty, black-and-white, no room for error or interpretation.

"It's time," I declared to the internet at large in December 2014. "My smartphone has made me dumb. Its pull is too strong. I can no longer control the beast in my pocket, and I want my life back. BYE FELICIA. This invention will not take my attention. That is why I'm dumping you." I held on for less than one month.

My second offense was in 2018, after I wrote the following for all of Instagram to read: "There are now more phones than there are humans on earth. What was once considered a tool for connection is now a replacement for real human connection, tough conversations, and chance. Tomorrow there will be one less as I unplug and return to real life."

This try barely made it off the ground. I called it quits after a failed attempt without it to fill a prescription for my son. I felt wishy-washy, but motherhood trumped the social experiment. I swore I would do better next time.

My most recent pledge, smartly framed as a trial this time, went out in my *Folk Rebellion* newsletter on November 4, 2020: "In my failures of unplugging, I've learned not to be absolute. So, I'm calling this what it is—an experiment with the help of a dumb phone. Albeit a hopeful one. Tonight I, for the 3rd time in 6 years, will attempt to get rid of my iPhone. This direct line of addiction, anxiety, stress, unwanted acquaintances, trolls, escapism, unwellness, and garbage that automates my life in a negative way has no excuse . . . no, no NEED . . . in my pandemic life. I, quite literally, have no more excuses."

The breakup didn't go as planned. As is often the case when you are trying to find the right path forward for yourself—it's not always easy and clear. Sometimes you need to adjust until you find the exact right path—*for you*.

Good thing it was just an experiment. When you "fail" an experiment, you are actually making a discovery that puts you one step closer to your answer. I love this mindset for life! It's in the trials, the inquiry, and the curiosity that you find the truth—after many pivots.

Of course my concerns about the iPhone when it first came out were not the same as my worries now. With new technologies and innovations come new fears. When the TV was invented, people worried it would ruin their eyes; the telephone was suspected of being a conduit to evil spirits; people believed that trains carrying human bodies at thirty miles per hour would surely melt their skin. Hesitation around the unknown is fundamental to humanity's individual and collective safety—which is why the rapid universal adoption of all creations in this new frontier called the digital revolution is so concerning to those of us who have awakened from our blue-lit screens. While the initial concerns about many of these technological advances were somewhat irrational, the notion of caution is completely reasonable.

And yet when we hit peak adoption of iAnything, the concerns of governments, religious institutions, educators, parents, doctors, or scientists were silenced for almost the entirety of the two decades. It was mostly cheers, not jeers.

The digital revolution is the advancement of technology from analog, mechanical, and electronic machines to the digital devices and services available today. It heralded the start of the Information Age, making info more accessible through those networks. Though both emerged in the late twentieth century, they continue to grow and change today. The internet has given us limitless abilities, including the power to globalize everything from our communication to our businesses. Technology gets bigger—and more invasive and unavoidable—with every passing day. This digital wave has been climbing higher for decades, but what started as a ripple now may feel more like a tsunami with no crest in sight. The rise of social media from its early days of Friendster and Myspace paved the way for media giants like Facebook to not only exist—but dominate.

We barely notice the things we do daily that would have been inconceivable a decade ago: We ask a robot to tell us if we should bring an umbrella when we leave the house or order a personal driver to our front door in minutes without interacting with another human being.

This digital eruption has clearly changed—and challenged—the way businesses operate, the way people operate, and the relationship between human beings and our machines.

On the surface, it's phenomenal to live in an age like this. Technology makes the impossible possible, and yet in many ways, we're thinking smaller as our worlds shrink to the size of screens. These days we are hard-pressed to find something or someone more interesting than the pull of our smartphones.

My God, how "phones" have changed.

In second grade, a boy called my house and asked me to be his Valentine. When caller ID was invented, if we saw a friend's parent calling, we'd throw the phone like a hot potato. I used to carry quarters with me on all my bike rides and misfit adventures so I could call my friends and tell them where to meet up. On 9/11, I used a pay phone to call my mom and tell her I was safe. The operator connected us for a brief enough second for me to shout, "We're okay!" before being disconnected. Phones were central to my upbringing and early experience. And yet now they seem to have an entirely different function.

The cell phone can probably be blamed for the demise of a relationship or two. When it buzzes, I now feel anxiety when I used to gleefully think of who it might be on the other end. The news is constant, bad, and following us around on our phones. Texts have become go-to abbreviated forms of communication in which sending a funny GIF fills the void where a substantive discussion used to be. In our fragmented world run by phones filled with free information, I miss the time when websites were still pixelated.

I'm often called a nostalgic Luddite for espousing my love of all things analog. It was technically my job to think about how tech affects us both positively and negatively. I try quite hard to not be that scared-of-the-future type, but the more our technology (and the insidious influencers behind it) goes everywhere, the more I become concerned about the influences we have to push back against. It's around us, on us, in our homes and schools, watching, listening, and "learning" from us.

We are too tuned out to tune in and register our current situation. We've been lulled into submission. The social and humanitarian impact of all this tech needs to be looked at. Judging by the data alone, we have good reason to be concerned about tech's pervasiveness—Facebook has 2.9 billion users, and in 2011, Nielsen reported that Americans alone spent fifty-three billion minutes on Facebook each month. That's more than one hundred thousand years. In 2020, viewers spent one hundred billion hours watching gaming content on YouTube. Every minute, three hundred million hours of video are uploaded to that platform. Instagram is on track to reach 2.5 billion users. Every day, ninety-five million photos are uploaded to that network. TikTok has surpassed Twitter, Reddit, and Snapchat in monthly active users, reaching a valuation of $280 billion in August 2021. For comparison, Facebook is currently down a cool few hundred billion since TikTok's arrival. (That might be why, suddenly, your Meta Instagram and all its promotion of Reels started looking a lot like TikTok.) Attention was going elsewhere—and that meant the money was too. There are 6.65 billion smartphones, with 83.37 percent of the world's population owning one. The top users touch their phones more than 5,000 times a day. The rest of us clock in at 2,617 touches per day (though this research was in 2016, so our dependence is likely significantly higher given the increase in speed, number of apps, and pandemic downtime). Ninety percent of millennials sleep with their phones. And 74 percent of Gen Z members report spending *any* of their free time online.

While we weren't paying attention, our entire lives have been co-opted by the World Wide Web, which means time online has replaced other valuable activities and preoccupations that actually fulfill us. Silicon Valley has created an "attention economy," which treats human attention as a scarce commodity, and profits from it regardless of how it's harming us and what it's stealing in the process. Just like our other resources—land, water, air—we only have so much of it before we exhaust it. This time the "it" is us.

Around 2012, Jean Twenge, an American psychologist and author of *iGen,* started to notice sudden changes in the big national surveys—incidences of depressive symptoms and loneliness started to go up. Other sources—like national screening studies on depression and statistics on teen suicides—showed the same pattern, with increases after 2010–12. The Pew Center's data showed that the percentage of Americans owning a smartphone crossed 50 percent at the end of 2012. Twenge found that social media moved from optional to mandatory among teens in 2010. And many others found that young people who spent more time on social media were less happy, and more depressed. This was a suspicious pattern, namely, a sudden rise in mental health issues when smartphones and social media became ubiquitous, and a link between social media use and mental health problems.

This new attention economy is bankrupting us. It's so predictable in its capitalistic desires. The collective shrug from the founding fathers of technology is both concerning and unsurprising. They continue to pile it on, despite the research and warning signs, altering our moods and, ultimately, our society. Tristan Harris, former Google design ethicist and cofounder of the Center for Humane Technology, has been warning us for some time via TED Talks, podcasts, interviews, and Senate hearings that the sophistication of the techniques used to mine our attention is exponentially outpacing our ability to protect it. Can we truly fault them for not looking far enough into the future to predict the ripple effect on generations to come? Surely it was a different world back then. But the ripples are clear now—they *could* do something about it.

As I learned when I tried to annex my smartphone, technology is becoming a necessary condition of existence in regular society. I've tried to bob and weave. Unplug and build boundaries. Unsubscribe. Decline terms. Opt out. But its tentacles are far and wide. As much as I would like to combat my addiction by going cold turkey and fully discarding the addictive substance, it's not realistic today to get rid of the internet entirely. And boy have I tried.

So what can we do? If it's impractical to avoid tech altogether (or time travel to the eighties), how do we ever live intentionally and find that balance?

Well, we can begin by opening our eyes.

Tech is influencing how we live, feel, and behave, but most people have no idea that it's happening. We don't know the full repercussions of our always-on instant gratification under-the-influence culture, in part because we can't study it fast enough. And not everyone has the luxury of a family intervention or a retreat to a quiet place where they can detox. Since digital technology is always upgrading, updating, and shifting, what we study in one month can be (and usually is) outdated in the next. It's like trying to pin down a shadow.

The negative effects from misuse or overuse of technology and digital mediums come at an incredibly fast rate yet are often invisible. Even when we begin to suspect that there are issues and attempt to question or study them, the tech is upgraded, and we've moved on.

The dangers don't stop there.

As online comfort zones replace in-person interactions, those on the front line with our children and young adults—teachers, therapists, parents—are sounding the alarm. When faced with any normal reality—be it a date, a new job, a school classroom—social anxiety, fear of criticism, lack of self-confidence, and communication issues become the clear and present dangers our protected children are beginning to face as they grow up in the digital age. It makes sense: The more our kids move inward—engaging with social media social groups, scrolling curated feeds, staying in controlled spaces (like a group chat)—the more their self-perpetuating anxieties continually increase. If the out-of-doors, uncontrolled world outside them is not experienced, how can they fully develop as people?

Imagine having rarely spoken without premeditating what you were going to say. These digital generations are growing up thinking, *then* typing, *then* editing, *then* sending, and have been doing just that for *everything*—asking a friend to get together, talking to a teacher about a grade, saying what's up to a date, answering a co-worker's question. It's all premeditated. It's front-of-house impression management 24/7. Furthermore, the less we make contact in person, the less normal it feels, creating a ceaseless anxiety cycle. Speaking off the cuff without the safeguard, I'm sure, feels *pretty fucking terrifying.*

When our words and thoughts are filtered, though, we are never showing our genuine selves. Yes, you have to edit and use socially accepted cues to communicate well, but our authenticity was undeniable during our fevered phone calls with friends or in our shaky voice when standing in front of the teacher who was about to sink our GPA. Being unfiltered is freeing. It's freeing because we experience hardship or difficulty, and we survive. When we bind ourselves up to be palatable, likable, same as everyone else, it's like personality constipation.

It's sadly ironic that the mediums that were created and later positioned with the messaging of furthering connection and making our lives easier, have actually left us isolated and overwhelmed and dealing with all the real human ailments that come from that. But becoming aware of these repercussions is the first step.

Tim Berners-Lee, the creator of the World Wide Web, thinks it "defeatist and unimaginative to assume the web as we know it can't be changed for the better in the next 30 [years]. If we give up on building a better web now, then the web will not have failed us. We will have failed the web." Tim doesn't think *it is what it is*. We're essentially still in the first five minutes of the first day of the digital revolution—that is, if you look at its endless upward potential paired with its current exponential growth. So all is not lost.

In 2019, Google announced at their annual developer conference that they would be adding "digital well-being" to their core values. They said they believed that great technology should improve life, not distract from it. These were their claims: "We're dedicated to building technology that is truly helpful for everyone. We're creating tools and features that help people better understand their tech usage, focus on what matters most, disconnect when needed, and create healthy habits for the whole family. We're committed to giving everyone the tools they need to develop their own sense of digital well-being. So that life, not the technology in it, stays front and center." But like most wellness trends, they're selling more than they can actually deliver. These supposed shallow pivots and virtue signals are often for show, created to counteract bad press and to posture without the implementation of any actual changes or impactful tools. Following suit shortly thereafter, Apple declared they, too, would prioritize digital well-being, announcing a new series of controls that would allow users to monitor how much time they spent on devices, set time limits, control the distraction of notifications by grouping them to arrive at certain times, and control device usage for children.

Not so shockingly, when you lay the responsibility at the individual's feet versus holding the company accountable for their creations, both initiatives haven't done much to quell our addictions.

When I read about these well-being initiatives, I felt depressed by the uphill battle to make effective change in the world. I was anxious about all the talking out of both sides of their mouths as they said the right things but did the wrong things—that is, for example, stating they cared about our health and our future generations but creeping into schools with an aggressive race to arms in getting their iPad or their Chromebook into the hands of every student (new consumers) and simultaneously lowering age restrictions despite what studies were showing. And so for now we must handle these issues ourselves in our *Surface World* to protect our *Inner World* until a larger entity, with more systemic influence, steps up to make meaningful change on a larger scale in our *Outer Worlds*.

I was overwhelmed by the magnitude of this influence in our lives until I realized I actually did hold all the authority, power, and influence where it mattered most: inside my four walls, my home, my life, and my kids' world.

Moments in which I feel unmoored and overpowered are usually the signal, my internal alarm bells, that it's time to pause and reassess. When I recenter and refocus on the influence that matters most— I realize that *that* is within my control. I don't need to be exhausted trying to fix the world for my sons at all times. Sometimes I just need "our world" to be fixed. And if I couldn't completely cut ties with the internet, I could at least instill checks and balances in a real way, like by not relying heavily on screens and not being afraid my kids will be bored.

Ultimately, that's our one surefire play. We need to take matters (not shadows) into our own hands, in our own homes. Understanding that we *do* hold control, in the context of our own lives, is a step in balancing our intentions with the influential pushbacks of our world.

And it seems I wasn't the only one to realize that. From the ground up, changes in microcommunities are beginning to happen in homes, schools, and businesses across the country. Buzz terms like "digital detox" and the "slow living movement" have entered the zeitgeist, answering these concerns. Wifi-free travel destinations are some of the most sought-after locations in the world. True disconnection—going truly offline—has become the definition of luxury. Unplugged retreats, adult summer camps, and meditation clubs are now a large part of the ethos and lexicon of today. No-tech private schools, apps restricting phone use, and restaurants that ban phones are just some of the ways consumers are balancing their relationships with technology. The pro-verbial pendulum is swinging—or so I hope.

That's why, in November 2020, I announced the experiment to get off the iPhone juice. It was no longer a necessity in my pandemic life. I had no clients, no business, nothing that required a phone. Finally, I had no more excuses.

During that time, when I gave up my smartphone for an old-school flip, I chronicled my trials and tribulations in a newsletter (yes, I know, still digital). I shared the exhausting, emotional, and sometimes de-lightful journey of cutting the cord that binds. My live journal took readers on my trip from smartphone to dumb phone. Each newsletter elicited hundreds of responses. What was edgy or insane at the start of *Folk Rebellion* was now clearly mainstream.

So many people, so completely tired of their digital lives.

Despite being a two-time iPhone quitter and one of the Trojan horses in the digital well-being movement, my last foray demonstrated that even I had a thing or two to learn about the impact of parting ways with my phone, for better and worse.

First and foremost, I learned the difference between guzzling infor-mation and gaining knowledge. The latter is a fulfilling pursuit, an example of harnessing the internet for its higher purpose. The former can become an addictive compulsion. It turns out I was an addict. The information drug comes in many forms. An addict's vice of choice could look like watching stocks in real time, consuming news alerts to be the first to know, or binge scrolling through strangers on Insta-gram. My vice is that goddamned Google search bar.

The dumb phone only allowed calls and texts. It couldn't feed my information withdrawal. After twenty-four hours, my monkey brain could no longer hide. The thoughts came in, wavelike, battering the surf-less shore of my reprogrammed cognition. I began tracking a mental list of things to look up when I returned to my computer, which I was obviously still using. (I'm not a masochist!)

What time does the sun set?

Where did that girl from *The Queen's Gambit* come from?

Best place to volunteer.

Transmission rates outdoors with masks on.

2020 book to give as a gift.

Can I study neuroscience without a degree?

How to sail in Brooklyn.

The list took on a life of its own. Within hours, it had grown so long I had to abandon it completely so I could cook supper.

The dumb phone's most brilliant feature—the fact that it had none—allowed me to see I had a serious problem. It took a few days for my information-addicted mind to detox. Once it did, on the other side was clarity, creativity, and calm. All I had to do was stop looking shit up. All I had to do was accept that it was okay not to know.

The second gift from my dumb phone experiment came by way of quite literally looking at the world differently. I saw my surroundings through a new lens: a viewfinder. I've worked with digital boundaries and mindful tech enough to know that you can't create a vacuum where a device used to sit. I knew my dumb phone wouldn't fill the space and time my smartphone did. I needed to pick up a new hobby. Photography was once my creative medium of choice. In my teens and twenties, it was my excuse to see the world and then hide away and think about it while watching photographs come to life in trays under fingertips worn bare. I owned point-and-shoot cameras of every kind. It was my passion and, I now realize, my escape.

Every new technology puts someone or something out of business. My iPhone was no exception. It cannibalized my hobby and took the process out of my photography. It replaced my escape with quick outputs and a record of 68,683 bits and pixels. It robbed me of an entire decade of creative expression and a practice in patience. Realizing this broke my heart. I now see how much worse off I was for it.

Recognizing the importance of nurturing my passions, I dusted off a film camera for the experiment. I walked through Brooklyn seeing my home in a new light. Strangers became characters in the story-board of my stroll. Every step of the process required patience, pause, and determination. Loading. Composing. Processing. Developing. Printing. It was a master class in being unhurried. My dumb phone experiment exposed my need for a life-giving hobby away from a life-sucking screen. I could utilize my passion pursuit to limit time devoted to the digital world and release some of the hold it held over me.

And most important, I realized how much I missed audio connection. The dumb phone threw me into the "Can I just call you?" realm. More conversations, fewer texts. I noticed an immediate uptick in my mood.

I used text messaging the way it was intended: as efficient short-hand. Digital dispatches were not meant to replace conversations. When my brother asked, "Why would you get a phone that can't send pics, vids, or GIFs? What's the point?" I responded with one word: *talking*. I chatted with him twice in the next four weeks. That was twice more than the previous month.

Text on a screen does not give me the serotonin boost I get from hearing my siblings laugh on the other end. Audio changes everything. It's the difference between nerve-racking notifications that keep the relationships shallow and shorthand versus a heartfelt connection. (It makes me think of the iGen crowd with their high-text/low-serotonin.)

I was desperately committed to my no-smartphone pledge this time around. Failing meant losing a large piece of my identity and the re-spect of my family and large swaths of the internet. My now-husband is a full-time dumb phone user, and I had visions of Hays eventually following in my dumb phone footsteps.

The final lesson is one I didn't see coming. It's also responsible for my final offense: December 27, 2020. Final because I won't be declaring a divorce from my iPhone again in the future. It's a pledge I no longer desire to keep now that my experiments allowed me to find the right balance.

Before the Wall of Shame hung in the cupboard, there was Mom's Phone Tree: four well-worn pieces of paper, each a different color signifying a different child, plastered with names and phone numbers. These analog Rolodexes were her paper portal to the community at large. Parents, coaches, teachers, doctors, babysitters, carpool rosters—anyone who helped her kids grow.

It was my mother's support system, written down and within arm's reach from the wall phone with the extralong cord. Each year, the Phone Tree's roots grew stronger and deeper.

Today's Phone Tree barely resembles the highlighted, starred, and tattered piece of love hanging on the back of a cupboard door. It's hardly a nineties relic. It looks more like a combination of Google calendars, contacts, group text messages, sports apps, and the like. It looks like an iPhone.

As a mom to an eleven-year-old boy, I require a Phone Tree. When I made my latest pledge, I didn't know, let alone appreciate, the fact that I was carrying one in my pocket. I didn't make the connection until my experiment cut off my Phone Tree at its trunk.

A dumb phone is group-text friendly but has its limits. My life-raft parent thread was more than it could handle. As a result, my stream of meetups and playdates, doctor recommendations, and COVID contact trace alerts went dark. I missed memos about soccer practice cancellations and notices of fellow families' quarantining. My son's social life took a hit. We literally couldn't be present because my Phone Tree people didn't know where we were.

Oftentimes people think the ditching of my phone is a desire for less when really it is just a desire for more of the important things. I am a person who really shows up in real life. This means I am not a "glow face" when talking to my son. I'm not a multitasking mom at the playground. I used to be both of those things.

Sometimes there are competing needs, and even in those moments, it's possible to find a way through. It might not be perfect, but it's not about perfection. It's about not fully succumbing to one over the other and losing myself in the process, but rather in retrying and restructuring until I get the balance right. My approach is something I have built based around my values. It's a daily, if not hourly, practice. I still sometimes fail—or, better said, experiment.

Some may say it's a privilege to live this way—and it is. But I need to be very clear here. It wasn't gifted or bestowed upon me. I have fought tooth and nail for this intentional life. I've burned jobs, relationships, and networks to the ground to preserve my priorities. I've gone broke. I've stood alone. I've been judged, mocked, and ridiculed. I've lost friends, lost business, lost respect, but I never lost myself when I was living with intention. It has been the hardest thing to live with choices that were best for me, my values, my kids, my dreams, my desires, and my health but that went against the grain. I weighed the odds and played them well. I saw the risk versus reward on the push-backs of life's accepted norms and decided to take a gamble. In the long game, my experiments played out in my favor. The risk *was* the reward.

I did not want to send emails at dinner, so I found a boss who allowed for that: me. I did not want a partner who prioritized media over make-outs, so I left an old relationship and began a new one with those goals in mind. I did not want to fill my life and surround myself with passing virtual acquaintances at a surface level, so I protect my time and go deep with those who've earned it. It's fair to say that I am present.

And after almost a decade of trying, I was able to finally trim back my digital life, iPhone included. The dumb phone benefits were immense, but they came with a loss far greater than the built-in perks of simple living. I need our digital Phone Tree community that makes our real-life community so great. As always, nothing is wholly good or bad. There are no absolutes in real thought.

We must decide what we want from technology, and what parts of our humanity we want to safeguard and protect. Our up-and-coming generations know nothing different from a devices-everywhere always-on world. If we don't teach them to question the pros and cons of welcoming a new product into their lives, who will?

While the internet isn't going anywhere, that doesn't mean we can't do something about it. Consider the Seventh Generation Principle: an ancient belief held by Native Americans that for every decision, be it personal, governmental, or corporate, we must consider how it will affect our descendants seven generations into the future.

Recognizing that there are real unintended consequences of our creations might help us determine what we want to usher into the world, and how. Just as the printing press wasn't responsible for the witch hunts that killed hundreds of thousands of women, it was a catalyst, helping spread propaganda. The witch hunts would never have been as destructive as they were if not for printed writing. According to James Dewar, author of a paper published in 1998 by The Rand Corporation on parallels between the printing press and the "information superhighway," many advancements have unexpected repercussions, both unforeseeable and uncontrollable, that frequently outweigh the ones that were intended.

A leading figure in philosophy and in the field of scientific methodology during the period of transition from the Renaissance to the early modern era, Francis Bacon said, "We should note the force, effect, and consequences of inventions which are nowhere more conspicuous than those three which were unknown to the ancients, namely printing, gunpowder, and the compass. For these three have changed the appearance and state of the whole world."

We've hit peak digital overwhelm. Most everyone has begun to burn out and feel sick, lost, and alone, like I was all those years ago. The silver lining? Instead of chasing the news on our digital devices or feeding into the 24/7 hustle mindset, more and more people are choosing real life and downtime. Living IRL and analog are the new goals. But it's important to realize that short breaks like these aren't enough.

So where do we go from here? If these short breaks aren't enough, what are?

It's clear that balance and moderation will be key to protecting the fabric of our families and society. I finally learned the great lesson from our family's Wall of Shame. Life is not black-and-white. We must find room for gray.

My definition of gray looks a lot like another tactic from my mother's nineties parenting handbook. She locked the front door to our family home. Strangers, acquaintances, deliverymen—she largely ignored unexpected visitors. Instead, she kept the side door to the kitchen open for friends. They were welcome anytime and came and went as they pleased.

As I've returned to iPhone life, I've developed my own version of a side door for my closest friends: a phone call.

A study performed by biological anthropologist Leslie Seltzer at the University of Wisconsin–Madison discovered that hearing someone's voice over the phone can reduce listener tension and increase oxytocin (aka the love hormone). The effect of receiving a text message, even a highly mushy, lovey-dovey one, is diminished. A real voice has the ability to alter a person's biological response. Think back to a moment when you had a chuckle with a friend over the phone and how that experience influenced your disposition. Exactly how long has it been since an LOL accomplished that? We can use these tools to be more life-giving, less life sucking once we recognize what we are trading.

My community can get ahold of me just by walking through anytime they like. And those who ring the front doorbell, so to speak— the DMs, @s, emails, and other stuff from the internet—can wait. They're not on my Phone Tree and don't belong inside my house.

After all, it's not that tech is necessarily evil; it's that our relationship to it is unhealthy, especially as those with influence work to pull us in. I love the internet. Hell, I was addicted to it. It's given me new friends, a business, a global communication platform. But when push came to shove, the internet needed to go back to what it was meant to be: a utility, not a replacement for human connection.

It's not just about digital detox. In its full form, I now see it's about an influence detox. Just the capitalistic *idea* that we're supposed to be constantly available is the root of the trouble. It's not the tech itself, but the influence of the notion of it, how we use it, and the expectations that come with it.

Not long after my iPhone came back on the scene, my son swatted it, while declaring, "No phones in bed, Mama!" It was 7:30 A.M., and I'd already broken three of my own rules: no phone in bed, no phone before 9:00 A.M., no phone around Hays.

Masterfully failing.

To him, it didn't matter that I had a good reason or something that "couldn't wait." These were the rules, and I didn't follow them. Period.

Not only is technology complicated—and that's putting it mildly— it's also thorny to talk about. Bring up this very polarizing (yet needed and sometimes fun!) conversation at a party, and you're likely to hear about the effects on our love life, family life, leisure time, and at work. On one end, you'll have those who are questioning and a bit fearful of where we're headed, and on the other, you'll have those who offer their unwavering support for anything innovative and "disruptive."

Maybe it's hard to talk about because it's more philosophical than our internet-addled brains are accustomed to. Tech and innovation touch everything and affect all of us. Maybe it's because admitting the enormity of the problem is overwhelming and leaves us shrugging our shoulders. *It is what it is.* In discussions around it, you're likely to encounter topics pondered by the great historical thinkers: reality, existence, morality, theory, rationality, metaphysics, and science.

I've come to find out it's also really fucking hard to write about too. My sons and I are learning as we go. Having the conversations is more important than not having them. And asking the questions is way more important than having the answers.

Folklore   The world is undeniably digital.

Folktale   I need to be an adopter of all new tech to not be left behind.

Folk Rebellion   Technology should consist of tools that work for me, not be replacements for real life, and I should think intentionally before blindly adopting the next new thing just because everyone else is.

*Raise Hell   Start to look behind your screens. Who is pulling your levers? If you were able to pause it all, what would you see in its absence? What have you missed? Think about ways in which you can balance the return of control with the uncontrollable digital world. Start with the shift back from 2D to 3D. What real-life things would breathe real life into your life?

MODERN BULLSHIT

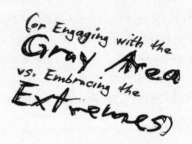

(or Engaging with the **Gray Area** vs. Embracing the **Extremes**)

Lately, I've noticed a trend in my book writing practice. If I spend any time at all on the crackling hellfire known as Twitter beforehand, I waste the rest of my day writing and then deleting, hyperconscious about what I'll eventually be presenting to the world.

Write. Delete. Write. Delete.

I keep reediting, reformatting, and revising the angle. I try to "bury the lede," as they say in journalism, which is usually a bad thing because it means trying to hide my point rather than being direct with it. Instead of saying what I actually mean, I start hoping that added nuance will allow me to communicate what I want without inciting someone else's rage. I try to avoid creating a succinct line that can easily be copied and tweeted by someone who doesn't like my opinion. Because of the nature of our culture today, I sanitize my thoughts, running them through a sensitivity test in my mind to determine who might come for me with pitchforks.

Not surprisingly, this sterilized version of my thoughts ends up being total shit, wringing the life out of my thesis, one placid sentence at a time. You're not writing what you actually need to say. Because when you cater to the haters, you're allowing the pushback to push you into submission.

It's safe. It lacks character. It lacks soul.

In 1990, more than thirty years before he was viciously attacked on stage at a conference for refugee writers seeking asylum, Salman Rushdie—an icon of free speech—wrote, "Human beings shape their futures by arguing and challenging and questioning and saying the unsayable; not by bowing the knee whether to gods or to men." We are at our best when we test boundaries and are brave enough to be ourselves.

There have been periods of time in my life when I have retreated. I consciously disappeared from the digital world and returned to my real one for the benefit of connection, quality time, and my mental health. But there have been times when the retreat was driven more by fear of the digital universe itself.

In the past few years, I've noticed that I'm not alone. I've started to see some of my favorite writers and thinkers shrink away, heading into the unknown too. Newsletters have stopped arriving. Essays are fewer and further between. Many of the artists who have remained are sometimes motivated more by attention seeking and algorithm chasing than by true inspiration, dreaming up the most viral creation versus the most salient, sensitive, or dimensional. Once again, the most successful ideas are not necessarily the best.

During the pandemic, we were forced into a retreat, whether we wanted it or not. That slowdown of our modern activity in the world is now termed the *anthropause*. When we humans went inside, nature began reclaiming our cities, creating an urban wild where empty streets became places to graze, play, and hunt again. Coyotes skulked in downtown San Francisco, goats traversed the highways of Istanbul, monkeys, wild boar, horses, and sea lions were suddenly spotted in residential areas, and rats roamed New York City's abandoned streets. But ultimately, it turned out that even the wild was dependent on our human activity. Without people to provide food through creating waste, growing crops, or bustling tourism—some species struggled. Even those frolicking rats eventually turned to eating one another, since there was no garbage on which to survive.

Our global slowdown didn't just affect nature—it also created an opportunity, a pause, for people to reassess. Many of us found ourselves rethinking everything—where we live, how we spend our time, our choice of careers. That moment of perspective resulted in a flight from offices, professions, and cities. But in the creative world of the internet, I also noticed many of the primary people I followed began to recede. One by one, they disappeared—the voices of reason, the moral compasses, the philosophers.

This retreat had already begun because of the emergence of cancel culture online, but when everyone went virtual because of the pandemic, the phenomenon became even more pronounced. People grew angrier and more frustrated with the state of the world, and what was shared and said on the internet became the only focus. Real life barely existed. So people hid.

When your "job" is to feel *then* share, both can become too much. I know this from personal experience. I understand the need to retreat from the arena, fold the cards, switch careers, live offline. I also recognize that if we're all busy navel-gazing, we can't contribute to creating community or rise up against the pushback collectively. But I can't help but feel concerned for what we're left with if all the thinkers, empaths, storytellers, truth sayers, and moral compasses retreat from the internet. What happens when they all proofread their words for pitchfork avoidance and keep their art tucked away in case it challenges accepted ways of thinking? I believe that now more than ever we need the people who recognize and call attention to nuance. The problems we face today often feel insurmountable. And as our culture seeks to correct its legitimate past wrongs, the pendulum can swing too far.

As humans, we naturally categorize what we see and experience into distinct buckets to make them easier to process—right versus wrong, black versus white, yes versus no. But the answer most often lies somewhere in between, in the gray. Data streams, metrics, and stats help us wrap our minds around our issues to a point, but without factoring in emotions, context, backstories, and the often overlooked difference between feelings and facts, we can't truly begin to fix anything. We need to understand what's influencing our human behaviors—the structures at play—before we can judge a situation. In order to do this, we need to be free to share even unpopular ideas. And when we do, instead of being attacked and written off, we need to be able to talk through these differing perspectives and biased lenses in hopes to correct what's wrong.

Without space for nuance and differing viewpoints in our society, New York City's cannibalistic rats seem—very darkly to me—like a great metaphor for our current downward spiral. Starved for valuable conversation and thoughtful analysis, left out to dry in a wasteland where everyone is competing to be the biggest victim, schmuck, or superstar—we have begun to turn on ourselves. The spirit of a place hinges on who is—or is not—there. Virtual ones, it seems, especially.

Currently we're living in a world of extremes, especially online. Of course it's easier to judge one another from the safety of our own homes, hidden behind screens instead of face-to-face. The arena is now a battleground, and similar to the spectacles in early medieval times, it's a center for ruthless conflict, public humiliation, witch hunts, public executions, and performed dramas. In the midst of this unchecked chaos, the most depraved have remained to watch and participate, and those who are turned off, who feel horror instead of glee at the extremes of what they're witnessing, have retreated to the shadows.

In their absence, the art, opinions, ideas, and thoughts don't entirely disappear from the modern-day virtual coliseum, but they grow quieter. Avoiding useless conflict, a true losing battle, they whisper in safer spaces, which become smaller and harder to find. This is not a good thing. It's grim, in fact.

Over coffee with an old friend in 2018, a former fellow bartender from Red Rock, I caught myself lowering my voice. We were talking about a hot topic of the moment, one sweeping the virtual coliseum in a wave of polarizing debates. It was perfect fodder for hot-take, clickable headlines. The only twist was that on one side of the argument, people had chosen "gray area" as their convenient weapon of choice.

Since I'd changed the nature of my lifestyle, I typically turned away from that kind of celebrity gossip in general, but the subject at hand was Aziz Ansari, a well-known and much-loved actor and comedian who also happened to be a voice I looked to for poignant commentary on our modern world. It seemed to me that we had shared a lot of the same perspectives, and he amplified ideas that resonated with me. In fact, we had decided simultaneously to go offline because the digital world was making monsters out of all of us. We both scrapped our iPhones for dumb phones and began creating platforms to discuss our respective wake-up calls.

Unfortunately, Ansari was leading that week's news cycle—and not for championing the art of unplugging.

In twenty-four hours, he'd gone from *golden boy* to *canceled,* at least according to reactive commenters online (today's version of a Greek chorus). That's how fast our churn-and-burn, eat-or-be-eaten civilization now operates. The arena might have been packed with an audience ready to observe and join debates over sexual assault, inappropriate behavior, boys-will-be-boys culture, and frat feminism, but really, above all else, this was about selling front row seats for profit. Who could hawk the most seats to a modern-day public execution?

As proof and point, the article that started the furor had a headline designed with one purpose in mind—go viral:

"I went on a date with Aziz Ansari. It turned into the worst night of my life"

Mission accomplished.

The story, on the now-defunct website Babe.net, was a three-thousand-word play-by-play of one night, told from the perspective of an anonymous twenty-three-year-old Brooklyn woman who'd had an encounter with the actor. The additional sources, fact checks, and context were scrutinized by those in traditional journalism while debate over whether this account of a gray area encounter warranted publication—and yet his life was blown up in a matter of minutes. Babe.net dropped the bait and the hordes pounced, becoming their own arena show.

Was Babe.net a news organization?

Well, let's break down the behind the scenes to find out, by looking at the sources, how they operate, and what their values and practices were before we even get into assessing the actual story against our own values.

Before it was bought out by a parent company in 2020, Tab Media was positioned as a global media start-up for people under twenty-five. It was started by Cambridge graduates George Marangos-Gilks and CEO Jack Rivlin in 2009 as a tabloid site in opposition to the university's existing newspaper. The umbrella company went on to publish The Tab and Babe, content farms both modeled after predecessors like the Bleacher Report and HuffPost. They relied predominantly on unpaid and/or underpaid writers, willing to barter their work for a first byline at the new twentysomething cool-girl media brand of the moment. And because, in our late-stage-capitalism world, people prize youth above all else, the scrappy start-up founders of a *new* generation traded their young audience, young know-how, young reporters, and young swagger for big money. From who? The *old* guard, the dirt-slinging dinosaurs in fear of being replaced, of course!

In the same way I once used my youth to pave my way into the digital universe as a liaison between C-suite executives and the new modern world they didn't fully grasp, Rivlin—the hustling, elite-school-bred CEO of Tab Media—got a door opened into a meeting with Rupert Murdoch because—do I need to say it? He knew a guy.

As the story goes, at least as framed by its founders to the press, the future CEO, Rivlin, showed up in London hungover, disheveled, with glitter in his hair from the night before, and wielding a book of viral headlines and possible exclusives for his millennial outlet. Their hook—what separated them from competitors—was that they understood their audience on a different level. After all, their writers were the same age as the target demo. In fact, the average age of the editorial staff was twenty-three. And the reporters were often even younger. In Rivlin's words, "If you put those guys in charge, it's chaotic and you get something pretty unvarnished. It also helps that they're always up for a scrap."

Always up for a scrap. A takedown. A polarizing hot take.

By fall 2017, Tab announced that it'd secured Murdoch's backing to the tune of six million dollars. This is the same Murdoch, of course, who is credited with creating the modern tabloid: newspapers that focused less on *news* and more on controversy, crime, and scandal.

The Ansari story went live on January 13, 2018, mere months after a second cash infusion—ten million dollars in funding from Murdoch and the venture capital firm Balderton Capital. This is important because it gives you a glimpse into the motivations behind the website and the influence a group of twenty-year-olds were under to run certain types of stories and deliver on their promise.

When you pitch yourself as the next "unicorn"—a term used in the venture capital world to describe a start-up company with a potential value of more than one billion dollars, you have to start performing for those big investors ASAP. You're under the influence of outside pressures associated with underperforming for your new bosses.

A bombshell report—like, for instance, a story that charges a celebrity with inappropriate sexual advances—can do a few things that are good for business. First, it obviously generates an increase in traffic to the website, attracting new followers and readers. This also allows the site to put advertisements in front of your eyeballs. The more eyeballs, the more profit.

Second, the true "bombshell" nature—in this case, the notable person involved and the headline, as opposed to the actual content of the report—garners a lot of press, which further feeds the machine of traffic-equals-eyeballs-equals-money. But it does another thing too: It creates brand awareness. Under the umbrella of Tab, Babe was trying to be the next Bleacher Report meets Vice but for young women, specifically aimed at "girls who don't give a fuck." Yes, that was their slogan.

If this upstart media company wanted the next "it kids" of the NYC media scene to work for them, they needed to invoke a comparable laid-back culture on the cutting edge. For that borrowed interest, Babe.net rented offices in Brooklyn, around the corner from Vice's headquarters. On their website, they announced their funding in a blog post written by the CEO. He said their inaugural Tab team was comprised of fifteen like-minded visionaries: friends who were crashing on couches, spending nights at the pub, and dealing with disciplinary actions. "Tab Media's audience is the generation of people born after 1990," he wrote, and "it's great to have the support of so many people born before 1960." They really leaned in to the generational divide for profit, brand, and influence.

An enterprising, hungry—and *yes*—young editor, a twenty-two-year-old recent college graduate according to *New York* magazine's platform The Cut, finally got the perfect golden nugget for a media start-up looking for conflict-stirring, traffic-growing content. Babe had been previously running a campaign soliciting stories about sexual misconduct for months in hopes that they could unearth something big—clickbait gold, especially in 2018, when #MeToo was heavily trending. She was under influences of her own. These types of stories were ruling the zeitgeist of the time. Now the speculators were editors like her, looking for that viral headline, their version of a gold rush.

Tab had an incentive-based structure, unusual for press, but similar to many sales jobs. The more you produced, the more you earned. Babe.net offered "prize money" for hitting specific page view goals.

I've had enough low-paying, commission-based, tips-only, stepping-stone jobs myself to know that they make you hustle. If your basic needs aren't being met with the help of a decent salary, you will do whatever you can to get that cash for rent, food, or transportation to that potential opportunity or potentially exploitive job. Perfect for the young, scrappy, driven, or desperate. Practices like having unpaid interns and salespeople are now frowned upon, but they weren't when I was coming up, so competition was fierce. But where there's a will, there's most definitely a way. As opposed to calling it a publication, the people at Babe.net specifically referred to themselves as "between a platform and a publisher," which allowed them to justify the unpaid labor. As a platform, they could argue that their contributors were just willingly posting content for content's sake, like on Facebook.

So the conclusion can be drawn that Babe.net's writers and editors who are responsible for what goes out onto the influential megaphone of a giant website, can only pay the bills by having the *most* clicked headlines. They will create under the influence of their needs and their knowledge of how to win in the coliseum—which is of course (as you now know) based on the influence of the culture they learned it from— that inciting rageful online conversations equals viral success. And paying the rent.

Data, testing, science, and tech people guide most mass-market publications' decisions about everything from the length of articles and videos to what tone to strike and types of content to prioritize. That's who and what really determine what will catch and keep your attention, for how long, and what might encourage you to share. Why, you ask? Well, to get the most eyeballs on it and sell more of those pesky ads again, which is ultimately about getting the most money out of it and you, of course. This is a perfect example of when the metric is not actually about being the best but rather is about being the best at selling or promoting.

This demonstration of looking for nuance might feel long, drawn out, and a bit much, but it's how it works in practice for me when analyzing my influences. If done well, it generally makes me more aware of the influences at play, less easily tricked, and open to the middle-of-the-road discussions rather than falling for the polarizing hot takes. Bear with me.

Babe.net and that vertical's editor got what they wanted out of the Aziz Ansari story. The post garnered 2.5 million views in twenty-four hours and was covered everywhere—from *The New York Times, The New Yorker,* and *The Atlantic* to TMZ, *Variety,* and CNN—and was critiqued, scrutinized, cheered, jeered, and retweeted about by, well, just about everyone. It caught the attention of the zeitgeist and was "the thing" for a very noisy time. Sides were taken—was this a bad date or was it bad journalism? Was it assault or was it predatory behavior? Was he an actual predator or just mildly (more than mildly?) skeezy? Was it an example of the pendulum swinging too far— cannibalizing the #MeToo movement and actual victims of sexual assault—or was this what the movement was about? Was this story something that should've remained private or was it fair game? Regardless of which side you were on (or *because* of which side you were on), it was feeding the machine. Different genders, different generations, and even different waves of feminists feasted on one another online, essentially putting more cash in Rupert's pockets.

When I lowered my voice over coffee, I was having a heated conversation with a friend about how I thought the whole thing was just wrong on so many levels that I couldn't even begin to dissect it (though I would try). Meanwhile, the young people at the table next to us were also speaking on the subject but in a brazen, unabashed tone I'd become used to associating with someone in a generation younger than mine. To them, this was right versus wrong, black versus white. All in stark contrast. As I eavesdropped, I realized that we agreed on fundamentals: Men should not behave in that manner and intimate encounters should not feel uncomfortable. But the similarities stopped there because, to them, any assertive and insistent behavior in the bedroom equaled assault, which is what the "news" website had implicitly suggested in the way they communicated their one-dimensional accusation. To them, Babe.net *was* a destination for news.

I might have quieted my voice at the time. But now I'll dare to utter the words I felt influenced to whisper back then: Not all sins are created equal, not all information is valid, not all intentions come from the right motivation (and vice versa), and while callout culture can force gray area conversations around accountability, the influence of social contagion and groupthink needs to be dissected and challenged, even when it makes us squirm.

It's incredibly important to redefine sexual dynamics in a culture that's been under the influence of men since the dawn of time. Labeling anything black or white by leaving just enough gray is doing a disservice to both genders. I hate to pick on the Lean In movement again but it is a great example of being *so* unknowingly under an influence, that in a quest to disrupt the status quo, you unknowingly perpetuate it. The attempt to make the pendulum swing by helping women get a seat at the table with men missed the mark completely because the solution posited was to *behave* as a man would, *within* that existing patriarchal structure instead of to tear it down. Frat feminism and hookup culture are of the same ilk. The idea is to create a pendulum swing, but instead of owning their own impulses, girls attempt to behave the way (sometimes deplorable) boys always have, in a structure that was created by men long ago. In either event, women hide their true selves and authentic feelings in order to pass. Not surprisingly women/girls posturing as men/boys hasn't worked out so well for us—or them. Either way, we are again trying to *force ourselves into existing structures* built on generations of influence rather than *adjusting the structures themselves.* The true solutions that mend the disconnect—whether they be about intimacy, communication, or enthusiastic consent to the temperature of an office thermostat, paid maternity leave, or work-life balance—happen in the gray.

When megaphones of influence impact a generation of its young readers, as in conflating a bad date and messy misogynistic sexual behaviors with sexual assault, we are in danger of false equivalences and false influences. We must wade into the messy, murky, complicated waters of gray area conversations, even if the weapons of polarizing absolutes make us want to retreat.

Today a fumble or an unpopular opinion can mean losing it all. When family, relationships, reputation, finances, safety, and privacy are all at stake, speaking nuances aloud may feel like too much to lose. The risk may not feel worth the reward.

I'm not a very risk averse person. That's probably obvious at this point.

Getting in the ring of life has always required huge personal risk. But being outspoken in our modern-day coliseum is even riskier today. We have all been influenced by this.

Erasing things—words, books, people—may seem like an easy way to right the wrongs, but it's not a strategy that helps create meaningful discourse and change. It doesn't shift influence. The only thing that does that is talking about and learning from mistakes. By doing that, we can go beyond just standing up for our own selves when it comes to influence, and start imparting what we value as good influence. To do that, we must engage in debate and be open to nuance—and today, that's a giant risk, making it a conundrum of epic proportions!

I'm forever lamenting that if we don't teach our young to think for themselves and use nuanced analysis to consider ideas, they won't know any difference. What I really mean is that young people, like my sons, will take ideas at face value and for granted without understanding that there's more than one way to view an issue. They won't understand that change is possible with rigorous, real conversation.

We need to consider context, and influences of the past, before we simply disappear things. If we censor and shelve books written through the lens of a different time, or cast aside movies, culture, statues, and people because the content doesn't align with our values or offends by today's standards, the next generations will have nothing from which to learn. If we remove all triggers, prioritizing every sensitivity, they won't be capable of sitting in uncomfortableness or have the resilience and perspective to observe something and assess what is and isn't valuable about it.

We can hold two ideas in our minds at once: We can be critical of how it was, or how it is, and simultaneously prideful about how far we've come while remaining resolute in how much further we must go. We can look at the foundations, structures, and influences that allowed a given incident to happen in the first place, tracing the route of how we arrived to now, and make sure it doesn't happen again. And yes, this can apply to the internet, influence, and gender dynamics but also among a million other arenas too.

History is a living lesson. Without examining it, we lose our most important guidelines for creating a better world, for not replicating the same mistakes. What we don't fear, we don't protect. In essence, turning a blind eye to our past makes us take our present for granted.

Before we had an online archive of, well, everything, mistakes lived in the past, often quickly forgotten after an apology. But cultures shift, times change, and people (thankfully) adapt. Mistakes and misunderstandings happen (along with some egregious wrongdoings), and sometimes they're forgiven. It's just that today, our technologies, our digital bread crumbs of indiscretions (even teenage ones), never disappear.

Nuance does not translate on social media or in viral headlines. It doesn't sell. For example, differentiating between a Harvey Weinstein and an Aziz Ansari isn't as interesting to amped-up hordes as burying them both. It's more work to look closer. But if we allow all our positive influences to retreat in distaste, and if we replace our gray areas with polarizing diametric opposites, what's left is usually only the influences with something to gain. In the same way, we must protect parts of our history in order to learn from them—we have to return to the coliseum of our modern world to fight the good fight for exchanges of actual ideas.

As the character Baal in Salman Rushdie's *The Satanic Verses* phrases it: "A poet's work is to name the unnameable, to point at frauds, to take sides, start arguments, shape the world and stop it from going to sleep."

The silver lining of the rampant fear of being ostracized for saying something unpopular is that many of the artists who retreated for fear of being canceled created deeper work while hibernating. And now we're starting to see it.

Ansari got his ass kicked in the coliseum. He withdrew for a bit. He seemed to have learned a lot, or so we hope. And then he got back into the theater.

"I don't know, man," he said in his comedy special after the fact, which was more like a performance art piece (fewer laughs, more message). "We just gotta figure out some way to have some empathy. We're all kinda just trapped in our own little world. And unless we figure out how to talk to each other in real life again, it doesn't matter what the problem is. I don't know what the answer is. Maybe click on some of the stuff they click on for a few days. See what's going on."

Lately I find myself once again looking to him. On one hand, I find his behavior unacceptable. On the other, I'm aware of the cultural influences on him as much as on, for instance, that young editor at Babe.net. You can use the right language but not do the right thing. When Ansari appeared as the woke poster boy up front and the boys-will-be-boys misogynist in the back, it felt hypocritical for many—his date especially. And when the writer of the Babe.net article claimed feminism as a reason to champion her story, but called an older female journalist who criticized her work "someone nobody under the age of 45 has ever heard of," making fun of her appearance and age and la-beling her a "burgundy lipstick, bad highlights, second-wave femi-nist," it made it appear like she was virtue signaling or unaware of what it meant to actually champion women's rights and feminism. I don't condone his, or her, actions. I do think people deserve second chances. Because making a mistake doesn't sap you of all your value.

If we all welcomed opportunities to engage in challenging in-depth conversations à la Ansari, we could in turn become good influences, or good spokespeople, for things that need to be discussed. Exploring nuance as part of our ongoing endeavor to understand what's shaping the influences around us would give us the courage to say the things only whispered in the wings, and not stand alone on a stage in a coli-seum fueled by fear.

Folklore   Everyone's saying it, so it must be true.

Folktale   If everyone's saying it and it must be true, I am to believe it.

Folk Rebellion   I understand the motivations of others, group-think, and I make my own decision before jumping to conclusions.

*Raise Hell   Think of a conversation you're meant to have, but you keep it to yourself. Is it something you shy away from for fear of the unknown? If your instinct or opinion isn't the same as others, what could be the benefit of wading into the messy waters of discussions without clear winners? If you have had the messy gray area conversations and they didn't go well, what could've been done better? Did they happen online, anonymously? Or across the table from a person over a cup of tea? Are there people in your world, in which you hold influence, who will lose out in some way by your staying on the sidelines of the coliseum?

80's MOM

(or the **Importance** of Weighing the Odds vs. Letting Fear Rule **Your Life**)

Stoop is life in Brooklyn. It's our dining room table for folded slices dripping with pepperoni grease, our local pub for good drinks on bad days, and our jungle gym substitute for kids without palatial suburban backyards and greener pastures.

So as soon as we moved into our current place in Brooklyn, we started spending time out there. I noticed a certain pattern right away. As my then five-year-old son played outside, well-intentioned neighbors would call out cautions and corrections:

"Honey, be careful!"

"Oh no, sweetie, don't do that!"

"Where is your mommy?"

In those first few days, "Mommy" was usually in the entryway putting on shoes or grabbing a tote. And, in the less than two minutes it took to lace them up or sling the straps over my shoulder, the locals had appeared as if from nowhere, aghast at seeing my child climbing the railing of our stoop—which I had instructed him to do. We'd just arrived the week prior, and I was still adjusting to this version of communal living.

Intentions, whether good or bad, can have a lasting influence.

I was a few years older than Hays was at the time, maybe seven or eight, when my younger brother and I rode our big wheels down the sidewalk to our local library, five doors down from where we lived. It had the perfect ramp for our very loud plastic lowriders. We were thrilled with this discovery, which turned our sullen moods around after being kicked out of the house by our mom. In true eighties fashion, she had rested a lit cigarette in the ashtray as she instructed us to "go play!," clicking off our cartoons and turning the dial to her daytime game show *The $100,000 Pyramid,* phone receiver tucked between chin and shoulder.

It was lost on us kids that libraries were a place for quiet. We were not quiet kids, and we'd just discovered a new game. But like something out of a bad eighties sitcom, an elderly librarian emerged from the building, shaking her fist to the heavens, and shouting at us to "beat it!" or she would call the cops. At least that's how my kid brain remembered it—a formative memory filtered through the influence of my favorite *Dennis the Menace* show. Though her intentions were less generous, she had an impact on me all the same.

I was the beneficiary of one of those idyllic childhoods people are nostalgic over today. I roamed, wandered freely from friend's house to friend's house, and rode my bike anywhere I wanted to go. If this were today, one might think I was a latchkey kid, a term to imply a child was unattended and therefore uncared for and in danger. It was often used as a put-down to shame parents who left their children unsupervised. But my mother was always there, and we were *greatly* cared for. We just didn't always have to be underfoot; in fact, we were encouraged to *go:* Go play with your friends, go outside, go but come back before dark, go be a kid. The independence of being left alone to my own devices outside of the constant purview of adults, and the trust my mother had for me (and the world), shaped much of my identity. Parents of the Gen X "latchkey generation" shouldn't be shamed for the way we were raised; I think my mother should be praised. She instilled a sense of freedom and unshakable ballsiness in each of us kids, allowing us later in life to push back against society's campaign of fear in everything from how we parent to the ways in which we make money and how we generally live our lives.

Of course my mom doesn't remember it that way. She is aghast when I remind her that she sometimes went for an emergency milk run to the local Byrne Dairy, leaving us four at home alone. She is in disbelief when my sister and I regale her with stories like the time we were riding bikes (a whopping full mile away!) from home and my sister crashed into a thornbush because she was too young to operate a bike properly. She is doubtful when I tell her I babysat a baby when I was twelve, even though I can point out the house. I have countless memories like these—building a dam to create a swimming hole in the ravine, snow days spent walking with saucers and cookie sheets to sled down the shopping center road, cutting through backyards to Grandma's house when we ran away, and being left in the car (the one where we never wore seatbelts) on a regular basis. But if you ask her about any of this, she "wouldn't dare!"—her stunned dismay is filtered through the lens of now.

Times have most certainly changed. Today the influence of false fear-based narratives runs the risk of ruling our lives. How we raise our children is one of the most drastic examples in my life because while I can turn away from living fearfully in my home, in my work, or in my personal life—with my kids it's more public. So there are more variables at play in the form of well-meaning people and new laws and legislation that are also under the influence of these fear-based narratives. Rising above the generally accepted parental norms of today and the pushback against it is something to *actually* fear.

The year prior when I was still living a suburban life, I was chatting on the phone with an old friend while driving. As I prepared to run into my local coffee shop, I turned around to my son in the back seat, and said to him, *"I'll be right back, OK?"* My friend was aghast. *"You're leaving him in the car?! You can't leave him in the car!"* I told her I could, that I had googled it just to be safe. She was afraid he'd be stolen, whereas I was afraid of having the cops called on me. At the time, I was in the clear. Today I would be arrested for child endangerment, based on a new law that was introduced in 2018 even though children are more likely to die within a parking lot, outside the car, than sitting within it. Rational decisions (and rational laws) are hard to make when under the influence of fear.

It's not that I have no fears as a parent. I do, but I temper that fear with the knowledge that we are as safe, if not safer, now than we were decades ago. Instead, my concern is more about what happens to a child who is never taught the benefits of solitude, daydreaming, boredom, watching the world go by, and being fearless.

How can I show him not to be afraid when the world is behaving otherwise?

When I turn to my risk versus reward internal meter (which is a thing I just made up), I try to consider how likely a negative outcome might be. Once I assess that, then I can think about what is lost on the flip side if I forgo the supposedly risky thing. At the time, knowing that my child had an almost nil chance of being injured by my leaving him in the car, but a huge loss of life quality if I tried to shield him all the time, I chose the former—even though it was no longer an accepted norm. Today it's not allowed at all. That's how quickly things are shifting. Five years later and my coffee run would've done more injury to my son by putting him through the governmental systems set up to protect him.

Maybe leaving a kid in a car while doing an errand seems like a more extreme example to some. Today our kids are monitored by baby cameras that sense breathing, movement, and sleep cycles, and, as they grow, are technologically tracked by GPS. Our playgrounds and parks have been stripped of seesaws and Bubble-Wrapped with rubber mats. So something as simple as climbing a railing on your own stoop is also now something to fear.

It's all part of the same thinking.

The top of the stoop is how far the practice of letting fear dictate our lives can reach—it'll reach every corner, every height, if we let it. That's actually the riskiest part: allowing ourselves to be at the mercy of people who are well-meaning but under the influence of fear.

We weren't always this scared. The increase in "information," real-time data, has made us fearful instead of fearless and well-informed. Though it's been proved that we have a better chance of winning the lottery than being bitten by a shark, our minds go all Lloyd Christmas from *Dumb and Dumber:* "So, you're saying there's a chance?" Add the branding of "Danger!" for the profit of news organizations—and our fears have been building steam ever since. Though it seems that the age of dread is suddenly everywhere and eroding our ability to make unemotional and rational decisions, the truth is that the seeds of influence were planted decades ago. And those seeds were sprouting, in fact, just as I was coming of age.

Google defines fearmongering as "a form of manipulation which causes fear by using exaggerated rumors of impending danger." Fear-mongering can make people nervous about the wrong things—like, say, dangers that are rare and unlikely—while *real* dangers get ignored.

The *unlikely* dangers that keep us up at night are often created by a *feeling* of irrational and widespread fear, known as a moral panic. In the 1980s, the news media was just starting to realize that fearmonger-ing equaled ratings. Media stories, politicians, and groups with an angle fuel the risks that might be real but are exaggerated or shared without context. Sedentary kids getting health ailments from staying inside (very common!) is way less interesting to the public than, say, kids stolen by strangers (not common at all!). Which one do you think makes the news?

One great example of false fear narratives and the trickle-down in-fluence they wield is how we indoctrinate the concept of "stranger danger" into schools, homes, and headlines, spreading mistrust and fear, when 99 percent of abducted children are taken by relatives. Of the 800,000 cases of missing kids in the United States every year, most were temporary (read: found), runaways, or a result of familial dis-putes. The instances of classic stranger danger abduction and/or sex trafficking accounted for only 115 cases. Total. Nationwide. Or put another way, one in 10,000 cases could be "stereotypical" *Dateline* kidnapping episodes.

Now, one might think *I don't want to be the one in 10,000. I will do anything to protect my kids from being that one.* I get it. I do. I don't want that either. No parent does.

But probabilities help you assess risk, so you don't make arbitrary decisions.

In 2021, ten years after my son Hays was born, I gave birth to an-other beautiful baby boy we named Beau. He just happened to be one of the two out of ten thousand babies born per year with a congenital heart defect known as hypoplastic left heart syndrome. As my doctor put it, I was struck by lightning. No amount of preparation or preven-tion could've stopped Beau from being one of the unlucky few. But we were lucky enough to be on the right side of medical advancement and, thankfully, today, after three very serious open-heart surgeries, Beau should be able to live a "normal life" (whatever that is!).

If I didn't understand before, I can certainly understand now how the notion of probability can go out the window and fears about even small risks can take over. Once you've seen things go badly against the odds, it's hard to invest in them. Who cares how unlikely a danger is when your child could succumb to it?

Sometimes we have to grapple with balancing "odds" against the realities faced.

To keep in the practice of making intentional choices, we must push back against the pushback of our world and of our own fears—learning when to stay resolute, and when to reassess. I still weigh the odds against risk/reward while trying to understand how the risky thing came to be perceived as more dangerous than the alternative. I consider all sides (and all influences) of the argument when it comes to considering the odds of the situation.

Was it outdated statistics? Was it built on gradualism—norms that became norms not because the facts matched but because that's just the way things were done? Was it not a protection for myself or my kids but rather a protection from liability on the other end? Was the protocol made because of one incident, like shoe bombs, bad lettuce, or tainted formula causing a moral panic from odds that were clearly in our favor? Was it a never-ending supply of links and headlines for profiteering media companies, or was it really real?

The first thing we told our family when we shared the news about Beau's heart defect was, "*Don't google it.*" We knew the internet was rife with terrifying statistics and outdated information about what we were going to embark on because those are the scary stories that most often get shared. Instead, we chose reputable sources with updated research, links of hope—children who have lived and now play soccer, play in a band, travel the world, marry, became doctors, and have children of their own. We offered support group information and as much scientific data as we could find about current advancements in treatment. We showed a brave face because we realized that if we could be brave, they would hide their fears too, which would have swallowed us up, had they not.

We had to believe because fear doesn't help.

Being protective of our children is a parent's job. When we come from a place of fear, our logical brain turns off. Feelings override rational thinking time and time again. That's why we have to push out the worry. Otherwise, we can't hear our guts. After considering all the influences at play, you ultimately can make an intentional choice in response to whatever it is you're supposed to be afraid of. For me, it's lower-stakes eighties parenting style and higher-stakes pushing back the fear of beating the odds of Beau's diagnosis. Whether it's offering a stoop to climb or feeding an NICU newborn when the standard protocol goes against that, it's an uphill battle for a mama who believes she knows what's best, despite all the well-intentioned people who would tell her otherwise.

Fear doesn't help us. But it's not surprising how much it rules our decision-making, when we look back at where we started—or more specifically, where I started. After all, the 1980s were a bottomless pit of alarmed panic thanks to the well-meaning Reagan era. Nancy Reagan's "Just Say No" antidrug campaign came out to do battle in the war on drugs with a parallel K-12 school initiative called D.A.R.E. The practice of featuring missing children on the side of milk cartons was started in 1984 by the National Center for Missing & Exploited Children. The same organization made popular—and later rescinded—the use of the term *stranger danger*. It was a perfect storm that had parents everywhere locking their doors, growing hypervigilant against lurking predators. Of course if the drugs were prescribed (also a growing epidemic) rather than sold on the street, no one paid any notice. You see the difference there? Real things to be panicked about (the developing opioid crisis) versus moral panic (the so-called war against drugs, which we now know was motivated more by racism than by any real risk posed to the public).

These initiatives brought extreme doubt about the world we lived in. No one could be trusted. Quicker than you could say no to drugs, parents moved their kids from playing out front in the neighborhood street into newly fenced-in backyards so they could monitor them as they played within the confines of a safe, partitioned environment. All this even though the one-in-a-million odds of being kidnapped are far outweighed by the risk of death for youth football players: one in

78,260. Being killed in an equestrian accident? One in 297,000. Lifetime odds of dying as a car passenger? One in 228. Or to put those odds into another perspective, you or your child are seven hundred times more likely to get into Harvard than be a victim of such an abduction.

Meanwhile, today online predators are a real threat and close to home. And yet we give our children devices with few checks and balances. These numbers are startling, and they are the type of odds that *do* keep me awake at night and my heels dug in for another year without a gaming console. I think of them as portals to predators, by which I mean, yes, actual sexual predators but also other types of new predators—alt-right miners mining for new, young, and impressionable (male) minds to join their angry-about-feminism gaming chat, ad companies, political movements, misinformation, or the platforms themselves. Numbers that should start a nationwide panic and demand for systemic change? One in nine children has been sexually harassed online, and one out of twelve children is sexually victimized through online chat rooms, social media, and gaming platforms. The places where kids are being kept "safe," indoors and on their couches, are, ironically, where they are most likely to run into predators. The *idea* of safety has trumped *actual* safety.

To me, it feels like pretty bad odds. I'll take the stranger danger any day.

The results of the eighties fear movement? Crime went down in the nineties, way down. Yay! Great! Who doesn't love less crime? But also, many argued, it was because of various cultural factors in the past that contributed to overall societal trends in the future. Systemically our *Outer Worlds* changed, altering our *Surface* and *Inner Worlds*. The influence of things outside of the fear movement—*real* things like structural changes and the trickle-down effect from them—were the true source of this change in crime stats, they claimed. As the seminal book *Freakonomics* examined, the passing of *Roe v. Wade* was the root.

The Donohue-Levitt study discovered that the people most likely to commit crimes were those who were born to people who did not want children or could not afford to care for them. With the advancement in legislation for those who did not want, could not afford, or could not care for babies now allowing them a choice, the cycle was broken.

Those potentially underserved babies who weren't born would've been coming around to the average age of criminality, eighteen, when Bill Clinton became president. Like the kid on a group project who did none of the work but still got an A plus, he reaped the benefits of such a structurally influential change. The eighties fear movement took some of the credit as well.

And now that movement has spawned another thing—the age of helicopter parenting—all based on a false pretense.

As a parent unconsciously breathing in the creeping normalcy of helicopter parenting, I was stopped dead by a 2014 *Atlantic* cover story titled "The Overprotected Kid," featuring a freckle-nosed, hesitant-looking, grade-school boy, wrapped in a pillow with knee and elbow pads and a helmet. He's holding his mother's hand out of frame and looking to her, it appears, for reassurance. The subhead had me slapping cash on the counter of my bodega—"New research shows he'll grow up more fearful and less creative." It's been my manual for raising my own son and now his brother, Beau, ever since. It helped me see where we were headed and adjust how I would raise Hays, who was two when I read it.

The chasm of difference between the past three decades of when I was raised and now, when I am raising my sons, is about as big as the ravine I used to climb, unsupervised. The parent-child relationship changes under the influences of our times. The norms begin to shift under the zeitgeist and what was once considered acceptable and normal—walking to school alone, playing in a front yard, being left home without supervision, riding a bike—can, and is, transitioning to being considered not just unacceptable but policed. Today's kids are safer than ever out in the world, and yet we are more overprotective than ever.

It might just be the American way. Like everything else, protecting ourselves is driven largely by big business and financial gain—liability from suing, products and gear needed for safety standards to protect from being sued, because in a society that doesn't have universal healthcare, somebody has to pay for the broken bone.

Despite this (or maybe because I refuse to be pushed into fearful submission), I made sure our most recent family vacation had a pool for Hays and friends to learn some important developmental lessons—frolicking, fighting, and figuring out risks on their own. Important memories and rites of dangerous passage took center stage—an accidental elbow to the face, hurt feelings, stubbed toes, bloodshot eyes, bee stings, and never-ending competitions with no winners—diving, belly flops, underwater breath holding. It was glorious and admittedly nerve-racking. I didn't want them to get hurt, but I didn't want them to think they couldn't do a backward belly flop because I kept shouting, "Be careful!" I didn't know if they could or couldn't, and neither did they. Which is precisely why they wanted to do it.

Curiosity breeds discovery. And I know discovery breeds self-confidence. If I am always telling my kids to be careful or stop them from following through on assessing a risk themselves, they will always look to me to be the harbinger of what they're capable of doing. And who am I to say if I've never let them try? I couldn't possibly know that answer.

Learning to take measured risks is a key component of kids' emotional and physical development. Without it, young people won't have past experiences on which to make more-informed (hopefully), smarter decisions. If we are so busy preventing accidents, uncomfortable situations, and pool parties, the valuable part of risk never happens. It's why I told Hays he could climb the railing on the stoop all those years ago. If he tries and successfully gets up there by himself, he's capable. It's been my rule since he was little. It allows him to figure out how to get where he wants to go and judge for himself what is too far. It teaches him to trust his own instincts. A few experiences of getting too high, stuck, scared, and unsaved, and guess what? He learned his limits. He adjusted his behavior. He didn't need Mom to tell him what was or wasn't possible. Will this translate to his first job? His ability to show up with confidence and make his own calls? Only time will tell, but I feel confident in his confidence. I know that every failure of mine has taught me about my capabilities and limitations. At a kid's level, that looks like climbing and falling, biking and scraping a knee, or being alone long enough for boredom to kick in to discover something new about their world or themselves.

If we allow the fears put into our heads by others to create fear in us, we will never live fearlessly. We must be the fearless ones who stand up and push back against a society that thinks it benefits from our doubts about our own instincts, or we will lose the game before even starting to play—by living in a world that underestimates us, and then acting as such.

What's truly frightening is the gradualism of our world's slide into a direction where, legally, you won't have the choice anymore to let your kids be kids. Sledding bans are happening in states across the country for fear of injury or lawsuits. Snow days are being replaced by virtual school days for fear of children being left to their own devices (pun intended) rather than their digital ones. While I might think it's a parenting crime to keep your children indoors all day on screens, I wouldn't call the police on the parents. Different strokes, different folks. Or as I say to my kids, different house, different rules—or values.

A tree in the shade grows less than a tree with a clear view of the sun. My baby trees are growing tall and strong in Brooklyn, overlooking the stoop they will one day outgrow. And just like them, even as an adult, I need the freedom to grow, stretch, bend, climb, fall, figure it out. We need clear views too.

In those times we seem overweary, crouched, protective, shuttered, stuck, maybe we need to pause, reflect, and check how high up the influences have climbed into our own psyches. The fear influences just might be the reason why we are feeling that way, and then we can look around, assess our situations for what they are *really,* play the odds of a misstep or two, watch out for the pushback of well-meaning people, and hold strong in our intentional decisions to not let others' overblown fears hold us, or our kids, back from that next step.

Folklore   The world is a dangerous place.

Folktale   I must protect against every risk that I am warned about, even if the odds are slim that this danger will affect me.

Folk Rebellion   I need to weigh the odds, gathering all the pertinent and true information in the context of the big picture of my overall well-being.

*Raise Hell When have you felt wasted worry in your life? What if instead you feared the loss of time spent being fearful? What would happen if you turned the tables and trusted in yourself again—and your neighbors? What if you practiced empathy instead of judgment? Could you be less afraid? If so, what could you achieve? Happiness, freedom, success, calm? What is the worst that could happen?

VULNERABILITY CLICKBAIT

(or Accepting the Needs of Your True Self vs. Drowning in the Motivations of Others)

In July of 2018, en route to a speaking engagement at that year's InfluenceHer conference (no, I am not kidding), I collapsed in Logan Airport. My passion and purpose as a digital well-being activist and work as a proverbial thought leader as the founder of *Folk Rebellion* had given me the ability to share my mission with many.

I thought it was my duty to make my message mainstream, to better the world, to help people. Yet there I was, the cliché I ran from my entire life. Clutching my chest, buckled over, thinking I was having a heart attack in a goddamned airport. When I landed back home, I went straight to urgent care and was diagnosed with costochondritis, inflammation of the cartilage in the sternum, brought on by (do I even need to say it?) *severe stress*. The stress of a start-up, of a semipublic life that was smashed to pieces behind closed doors, and of the influence of the professional expectations I was holding myself to (aka the *Folktales* I was telling myself).

For a month, I tried to get up and back to my life, but the pain in my chest—my literal heartache—put me back down again. Every time I thought I was better, I wasn't. But I had to get back to work, I told myself. *There are lives to change! Society needs me!* I felt I had to sell these ideas, promote myself, and educate the masses. To what end, though? Apparently (nearly) to mine.

And so I stopped. Purposefully this time because my body no longer gave me the choice of continuing to ignore my instincts. It was time for some serious inward reflection and a major permanent change.

I didn't think I needed to announce that.

I am not your clickbait. I am not your ecourse guru, your Instagrammable motivational quote, or your #goals inspo. I am not the second image of a before/after meme—*how it started, how it's going.* I am not your relationships-in-the-digital-age expert offering tips on "*how to consciously uncouple (and still get half of what's rightfully yours!*)." I am not a live-feed reality show for you to slow-eat-popcorn-gif while you laugh-cry-emoji my life falling apart.

And I am not what is, or is not, broadcast onto pixels and screens.

That last sentence is a way I ground myself, like a Post-it note on my bathroom mirror, but in my brain. I am *not that,* no matter how much the world wants me to think otherwise. Combating the ever-present influence of digital culture means a regular practice of intentionally pushing back against the norms of it.

Call it what you will: personal documentation, exposure, sharing, self-promotion, fame, content, visibility, optics, or transparency. Our urge to be public with absolutely everything in our lives is not only skewing our realities but making our lives untenable. The words *sharing* and *connection* existed before the internet and meant something completely different—something real.

But now we're starting to see the other side of technology's allure, the side that favors the constant stream of babble. Creators and users are no longer just voices shouting into the void. Their behaviors—often the bad ones—are being rewarded by the discover/explore/trending algorithm to influence other behaviors, and so on from there. It's a perpetual rendition of "this is the song that never ends" garbage you can't get out of your head or escape.

Regular people, brands, influencers, and others work to create content that's clickable and, even better, meme-able. And the most outrageous and envious usually win the algorithm game. But something new is running rampant on the internet as of late too. There's a fresh clickbait trend for which my midlife upheaval would've been a perfect fit: oversharing pain. It's like emotional prostitution masked as openness, bravery, and undiluted vulnerability—because it's most often wrapped up in attention, click, and profit seeking.

Vulnerability has officially landed *splat* in the center of the zeitgeist thanks to the master of it, Dr. Brené Brown. I am a fan. A big one. Her expert work and study on human emotions like courage, shame, empathy, leadership—and, yes, vulnerability—is profound both in what she's uncovered in her studies and in her ability to share it in a way that makes people pay attention. As a master of storytelling, she looks to these topics to better understand human connection and uses many mediums to share her research. Her books on bravery and risk and stepping into our power are bestsellers. She's earned herself a Netflix special, and her TED Talk has been viewed more than sixty million times. But in it, Brown pushed back. She has offered a much-needed course correction, one that people seem to be missing.

Vulnerability is not disclosure.

And there is no vulnerability without boundaries. "You don't measure vulnerability by the amount of disclosure," she says. "You measure it by the amount of courage to show up and be seen when you can't control the outcome." That's far different—and far braver—from the myopic trope of oversharing from the safety of our screens, which is what many on our slick social media platforms seem to believe. You see, like an old-fashioned game of telephone, but for the modern world, the internet has diluted the substance once again. When Brown's extensive research gets pull-quoted, slapped into a Canva design template, shared by a digital life coach, and pushed up the algorithm chain, a whole lot of people can end up with a misunderstanding of what it means to be vulnerable if that is their only context.

With even a quick glance on Instagram or TikTok, you're sure to see an endless stream of faces oversharing their lives to any strangers and "friends" who will listen, using words like *courageous, vulnerable,* and *scared*. Even on the professional, less personally based networking site LinkedIn, you'll find CEOs crying, opening their viral posts with lines like, "This will be the most vulnerable thing I'll ever share." Sure, it's brave to admit to the world that you're hurting by being candid about your life. But are these posts courageous or self-indulgent? When the desired result—their motivation—is an endless stream of direct messages, emojis, and returned oversharing—or, in other words, attention—it starts to transform into manipulation.

Knowing how the internet works, it's potentially also a distraction from what's actually ailing the person posting—IRL loneliness. Public vulnerability can have positive outcomes, of course, when it's authentic, when it's a cry for help, or when it's a genuine ask for community. Lives have been saved, organs donated, money raised, families reunited, jobs offered, and more. People do get help, and offer help, online. It's a wonderful thing when used with the best of intentions by people, and when it's used by the underserved, underheard, misrepresented, or those throwing out a last lifeline—a truly digital cry for help. This isn't the type of vulnerability I am referring to. I am pushing back on the cringeworthy performative sadness type of vulnerability that is under the influence of something else entirely—the kind that is providing fleeting shots of dopamine from each like, heart, or comment.

If you're sitting at home staring at your device and waiting for comfort or recognition—waiting to be truly *seen*—you could be waiting a lifetime. Sharing every feeling online, while isolated, turns *vulnerability* toxic. I can't help but wonder how much less lonely we would all be if we had those conversations in the real world instead. Some of my favorite people are people I met online—A "Hey! You like this, I like this!" DM by some extrovert on the other side of the country who I never would've known if it weren't for social media. I have nothing against the connections made through the tools of communication of our modern times. My issue lies in the superficiality of staying there. I became intentional with these tools, using them as they said—to connect—but pushed further, and away from those platforms, to develop my relationships outside of them. Connection isn't connection if it's just to connect. It's what you *do* with that connect, purposefully and thoughtfully, to breed connection.

Growing up, I was often shooed away from a table full of adults playing cards in my grandmother's home. It wasn't until later years that I learned the significance and weight of that table. Those gatherings were where family matters, financial matters, and business matters were discussed, handled, and kept private (read: secret). It's only today, in our overly connected digital world, that I can appreciate what those nightly rituals offered my old-school Irish family—real connection, nonjudgmental opinions, a gallery filled with those who earned the right to be there, and people who had each other's best interest at heart.

So I ask:

In a world that favors oversharing via avatars and has given humans the opportunity to become their own platforms, producers, creators, and broadcasters, what is to become of a forty-year-old mother and entrepreneur who exited a marriage and a business partnership in the same month? Who spent most of the past few years cobbling her life, identity, business, and strength together—privately, for the first time— away from the screens and behind the scenes?

The truth is, I have never been one to share with abandon. I can still hear my grandmother Peggy's voice in my ear. She was a stoic and kind first-generation Irish American. Her "pull yourself up by your bootstraps" approach to life trickled down and stuck with me from a very young age: *Go ahead and have a good cry* behind closed doors with close family or friends, but then, you put one foot in front of the other.

In Okinawa, that inner circle of nearest and dearest people is called your *moai*. It's a group of lifelong friends who meet for a common purpose to gossip, experience life, and share advice and even support each other financially when needed. It's these five-to-six-person group friendships, formed in childhood and lasting as long as ninety years, that researchers like longevity expert Dan Buettner believe contribute to making Okinawa a Blue Zone with one of the highest concentrations of centenarians in the world. The sometimes daily, most often weekly meeting is one of the most important cultural rituals for the Okinawan people, helping them live exponentially happier and longer lives.

If loneliness is as bad for you as smoking, and the internet is addictive and isolating, creating a kind of second family in the real world seems like a smart move. Now, years since this collapse and resulting digital detox, my social media feeds are shells of what they once were—I barely post, scroll, or follow anyone. I'm down to family and *actual* friends. It's noticeable. I get it. But how am I to "share," when all I've ever been taught is that private matters rely on privacy, exclusive of a social support group? In my mind, before I was influenced into thinking for a moment that sharing online would help, these kinds of issues were only discussed at a kitchen table over a game of gin rummy. So when I took a moment to pause and reflect on my core values, this is the mindset I came back to. I chose with intention to lean in to my earliest influences—my family—because those teachings were most in line with how I wanted to live my life going forward.

After my final reckoning with technological influence, my personal sea change included acknowledging that my relationship no longer worked either. Over the tremendously wicked years through the separation, divorce, business partnership dissolution, and brand closing, I received hundreds of digital messages filled with emojis and staccato sentences. And unsurprising to the New Yorkers, I've still got a list of people who texted me asking to be first in line for my apartment, and even furniture, should I not be able to keep it. *So sorry! Hope you're okay! How many bedrooms is your place again? If you have to move, I'll buy your basement setup. Would you take $250?* The finishing blow, when my divorce paperwork was finally finalized in 2019, I found out via text—a picture with the judge's signature and seal with a yellow thumbs-up emoji—while sitting on the toilet.

Kerry Lusignan, licensed mental health counselor, wrote, "Breaking up a family when children are involved is akin to pulling bones out of your body while you are simultaneously growing them."

Yup. That. The only words ever expressed, the words I could not find myself, to finally begin to describe what those years felt like. I had broken up my family, yes, and I had also broken up with my way of life. On every level, I needed to become a new version of myself, and I needed support in order to do that.

So was I really supposed to take the time out of grieving to create Instagrammable quote art to announce my life update? What happens when life gets real, and you really just . . . can't?

I'll tell you. You not only become nonexistent, but I quickly and uncomfortably found that the digital silence in my little corner of the vast internet led followers, friends, and foes to narrate and interpret the "lack of content" their own way. They made assumptions. I know because I make them too.

When I pushed back on all my extrinsic motivators that drove me to where I was (on the floor of an airport on the cusp of losing my mind), I intentionally dove headfirst into the cold unknown waters of my new off-the-grid life. Although purposeful, it was still shocking. In a single day, I turned on my out-of-office, put an "away message" up on social media, and ditched most text messages. It helped that we were headed into the holidays, and I had the privilege of being my own boss. It also helped that I was so broken and fed up, I didn't care what happened in my absence (a mess to this day I am still finding remnants of). I knew I'd feel some relief, a grounding, and slowness and presence. But an unexpected thing occurred—my instant disappearance pushed my *moai* to find me—offline, in the real world. My digital overwhelm and isolation were replaced by the calm of being surrounded by actual people. My people. And then people I didn't even *know* were my people— fellow soccer moms, faraway digital friends, other divorcées, an old client turned sister, my fellow hospitality industry riffraff, the bartender from my old local, and the ones who knew me "before." Before the marriage, the kids, the success, and the public persona. They were real people, flesh-and-blood humans, connecting. Old relationships and new. I had visitors, companions for meandering walks to nowhere, and an increasing number of stops for swing-by stoop sessions.

My speedy disappearance from pixels and posturing was more profound than anything I could find zipping under my thumb on the internet. I found that the world online is not the cold sweaty flesh of my best friend's hand as she goes through her cancer treatment. It's not the deep pools of brown in my son's almond-shaped eyes that I lose all concept of time in. It's not the legions of online "collaborators," but the one woman who stood by me and lifted me up professionally when I was ready to fold. The internet is not my mother's comforting hand, newly soft with age, which now feels like my grandmother's. I relearned the lesson that my followers were not necessarily my "real" friends. Creating consumable content out of life's highs, lows, and in-betweens isn't my creative calling. And the results of my creations and how they are digitally received (clicks, likes, subscribes) do not reflect my value as a human being (or the value of my work!). My vulnerability is not composed of prostituted emotions for exploitive trade on the internet; my vulnerability is in doing the hard thing without the watchful eyes and cheers from others. My bravery is not in posting a picture without makeup but rather is in opting out of the whole charade—purposefully.

These are the things I've taken with me. They are *my* reality.

And then, just like that, the internet (or at least an emissary of it) came back for me.

It was one of those bone-chilling days in February when a friend of a friend (or as you'd call them on Facebook, "my friend") approached me at a sceney networking space. It began slowly with a seemingly innocuous "What have you been up to?" followed by an implied slight. "Not really sure what you're up to these days or if you're still doing that speaker thing, but I've got a conference I think you would be great at . . ."

This was the off-the-cuff covertly patronizing comment that sent me down a modern age philosophical rabbit hole from which I haven't fully dug out.

The statement implied in so many words that I hadn't just given up tweeting and posting pictures but that I'd stopped working altogether. I already knew why, but I needed them to recognize that I recognized the not-okayness in their assumption.

"Do you think I am no longer working?"

"Oh! Well, I haven't seen any of your talks lately."

What they really meant was: "I haven't seen any of your talks on social media lately."

And there it was. No self-promotion, no existence. No pixel performances meant that I was not only a digital ghost but, apparently, an unemployed one as well.

On one end, you're grappling with doing what you need to do to care for your best self, while at the same time being pitted against what you need to do to still work or to be perceived well by others. It's so fucking hard to come to peace with the two forces pulling you in opposite directions.

But I finally did by recognizing and repeatedly reminding myself that the other forces are under other influences of their own. Their potential influences, and the ones they are under themselves, don't need to become mine. The only influence I needed to worry about was mine, and how I would positively influence those around me by remaining true to that intentional choice of not giving a fuck.

It's important to note that this conversation took place at the end of February 2019. My social media sabbatical was only fifty plus days in. I hadn't even disappeared for as long as it takes to "change your life!" per Charles Duhigg, author of *The Power of Habit*.

I wanted to scream at them (the *Surface World* influencer) and shake their shoulders.

"I am still here! But I was having a hard time! I am healing! I am falling in love! I am taking care of my son! I am trying to keep my apartment! I am thinking about my business! I am real, and I am a human with a life of experiences currently too real for a curated gallery! And my grandmother would be so mad if I turned my phoenix-from-the-ashes moment into bite-sized entertainment! It's only the stupid internet!"

When other influences are attempting to pull you in an opposite direction downstream—opposite yourself, your values, your goals, your plans—you can find peace, and strength, in your decisions despite criticism from others when you acknowledge the current of "norms" under which they are operating (drowning?). For every critique from someone under the influence, may you find confidence in your intentional choice to float above the current rather than be dragged down to do what the norms (and they) dictate.

So instead of shouting my feelings at them or giving in to the pressure of their pushback, I simply said, "I'm still alive."

I thought that particular status update was pretty self-evident.

Folklore   Life is meant to be shared.

Folktale   Speaking my truth into the void is healthy and normal.

Folk Rebellion   I understand that vulnerability does not mean publicly sharing my life and oversharing in a virtual world is not allowing me what I seek most: connection.

*Raise Hell   When you are tempted to overshare, check in to see what is sitting below that. What are your motivators? Is it attention? Or is it really connection? When you do share, are you getting the responses back that you desire? How do they make you feel? Are you succumbing to the influence of an algorithm that is profiting from your viral vulnerability?

Now draw a circle on a piece of paper and put inside it your "short list." Who are the people you would tell all the dirty details to? Who are the ones you don't sugarcoat for? Who would help you bury the body? On the outside of the circle, put other people you know and like, possibly love and respect, but who don't go in there. Can you cultivate a deeper connection with any of them to move them into the circle? It's okay that they're outside because you are going to draw a circle around them now too. If the inner circle is your *moai,* your outer circle contains your trusted connections. Okay, last part: Now write down the people who make you feel bad, doubt you, snub you, drain you, gaslight you, tease you, and steal your flame. Leave them on the outside in every way. Social media as well.

We are what we pretend to be, so we must be careful about what we pretend to be.

—*Kurt Vonnegut*

Part V
Influential

O kay, dear reader. It's your time to shine. We haven't come this far to completely skip over how to be a good influence. I'm not letting you off the hook that easily. Whereas all the previous parts were about the influence all around you (and largely how and why to avoid its pull) as well as the influential pushbacks to navigate around, this part is where you get to feel influence in your own body, in your own spheres, and learn how to recognize it, harness it, and wield it—for good.

Now, my intention up to this point was to show you all the ways in which influence can affect you as the receiver of it, as well as all the ways you could negatively affect others by being a bad influence. I know I've led you to believe that influence can be quite dastardly. While that remains true, I'd be remiss not to show you the flip side. In this section, we'll turn our attention to the good reasons for influence. And there are plenty! My hope is that with all your newfound skills, you'll choose to use your talents for good, not evil.

Our society needs more good eggs than ever. Stepping into your influential powers might just give the rest of us some of the inspiration, leaders, artists, philosophers, and general do-gooders we so desperately need. While up to now we explored how to think, in the following sections I challenge you to *grow*.

This part is going to show you how knowledge can and should be used for good to create change in the world. It took me a while to get here, but this book, and this section in particular, is an example of how I've done it. Those who don't want to fall under the influence of unhealthy societal norms often become influential to correct them, but as we discovered in Part IV, it's easy to get offtrack. Here, through my personal experiences, I'll show you the world of opportunity for good when ego, money, and fame are not a part of the equation. The positive side of influence can affect large shifts in societies and change and save lives.

You'll notice a switch within the following essays: There are no longer any underlying "alternative titles" like we had in the previous parts. Now that you've had your eyes opened to influence, you can see clearly with the transparency that your new worldview provides—you know how to pierce through the layers and seek out the deeper truth. Each title now gives you a straightforward action to take on the road from influencing one to influencing many. Your very first step, becoming aware of your own subconscious influence, encourages you to *wake up.* As you grow conscious of your powers to effect change in your own microcommunities (*Surface World*), the next essay urges you to *pick a fight* and change the narrative about something within them. And while you are hugely influential in your own right, when you find strength in numbers and *band together,* you'll be able to move the needle further or faster in the *Outer World.* Finally, you'll see how the biggest positive influences culturally operate with a "one to many" approach, hoping to *make trouble* and act selflessly for the greater good of all, making them hellraisers for restructuring the norms in the process.

From the inspirational influence of *Sesame Street* on young children to Ralph Nader's successful crusading mission to save lives with seatbelts to the impacts we can have ourselves on our homes, offices, communities, and children, you'll see that the opportunities to wield influence for good are many, and right at your fingertips. My hope is that you'll seek out these opportunities to inspire good in others.

You may be thinking (now that you're wholly awake and curious): What sets inspiration and influence apart? Fair question! Often the words are used interchangeably in our culture. *Who are your influences? What are your inspirations?*

The explanation that clicks best for me is to think of inspiration as *igniting a feeling.* Cynthia Morris, the thought leader behind *Original Impulse,* calls this experience an "emotional quickening." Imagine inspiration as a sensation of opening up our mind to fresh thinking, arousing our creativity, and expressing our soul. The sea can be just as inspiring as a TED Talk; the reward of a ripe tomato after the care someone took in their garden can be as inspiring as a piece of art. What inspires us is all very personal! But lots of good energy here! Influence, on the other hand, comes from a focused and often repeated force altering something. That force could be anything—the weather, your parents, your favorite brand—but its directed energy *ignites a change.*

There are so many things that are outdated in our culture. There are rules that have never been questioned, those that need to be broken, and questions that need to be asked. There are people who have to be held accountable, people who need to be inspired, and people who need more good influences and fewer bad.

In this part, I hope you can see how you can lead, or inspire, a good fight.

WAKE UP

My family's Friday-pizza-and-movie-night tradition is sacred. It's the day of the week when rules go out the window. You can eat in the living room, on the couch, in front of the TV, with soda, candy, and the TV volume at an unbearable decibel. There are only two rules that remain.

Rule one—we take turns picking both the pizza and the movie. And that matters because our tastes in Brooklyn pies run the gamut from flat, cheesy, and cheap to upside-down, deep, and delivered.

Rule two—you have to watch the movie even if it's something you don't want to watch, like, say, *Sing 2* for the eleventh time or one of Mom's dreaded "old movies." If it's important enough to the picker to pick it, then it's important that we pay attention. That means no second screens, scrolling, phone calls, reading, or leaving the couch. The taking turns, switching whose night it is to pick, keeps things fresh in our house.

Hays usually picks the new kids' movie of the moment, Beau gets classic Disney, Mikey likes a family friendly superhero flick or a documentary. Me? I focus on the outdated relics from the seventies, eighties, and nineties as time-traveling gateways to unexpected conversations. Because that's what happens when you watch stuff communally. A shared experience gives the participants something to think, talk, and bond over. And because (as we've already learned) we are what we consume, I choose this type of content to influence my kids to have nuanced conversations, waking them up to the fact that things change over time—often because of gray area discussions. Where some parents might choose to shield them from, say, the atrocities of 1980s misogynistic flicks, I like to put it right out there, and then talk about it.

That's why when Hays had a bunch of his friends over for a sleepover on a Friday, realizing it was my night to pick, he was crestfallen. Nope. I didn't give in. Rules are rules. I could try to explain that they'd be watching a movie from the past in order to learn lessons I found important about how some jokes don't age well, what the world was like before the internet, and how culture has shifted over the years. But they're kids. They don't care. Of course there was always a chance that Hays's mates might go home and say that our movie choice was bogus and they wouldn't want to come over next time. But I was willing to roll the dice. I've been doing this long enough to know that usually the exact opposite happens. When a person has only been exposed to one type of movie, sport, or even art, often anything different is uncomfortable at first—but it's also intriguing. Which is what I was banking on, even if it's slower paced, referencing a forgotten time.

In the vein of kid classics like *The Mighty Ducks* and *The Sandlot,* both favorites in our home, I chose *Little Giants* (dating back to 1994) as my pick, hoping the sports theme would hold their attention and Rick Moranis would hold mine. It tells the story of a group of kids who don't fit in their small town and find ways to overcome the odds and make their own rules. A perfect choice, I thought. Comedic, football, and pizza. Done.

It took longer than usual. They made it halfway before being offended.

> *Hey!* That's not fair! Why would they say that?! He's so rude! I would kick him in the nuts! She should kick him in the nuts!! They all should kick them in the nuts!!!

I was making popcorn when a pillow was thrown at the screen.

"*Guys!* Chill. What's up?" I asked, knowing full well what was up.

Scrambling over one another, one talking over the next, the boys all wanted to be first to express their outrage. It reminded me of when one of my siblings or I tried to be the first to get to one of our parents to tell our side of the story. To share the wrong that had been perpetrated.

Why would they make fun of girls being on the team? I don't get it.

This is what subconscious influences do for us and, oftentimes, to us. This moment made visible the result—the success—of all the hard work put forth by people with influence to begin to dismantle the gender biases of our world. These boys, at this time and this age specifically, were entrenched in their unconscious beliefs that anyone can play on a team.

I did not want to break the spell by telling these boys why the girls were being teased in the movie. The thought had never even entered their heads. Their core beliefs were not that *boys were better than girls* or that *females were equal to males*. Their core belief was that there was no difference between them. There was nothing to override or to overwrite. The instructions, patterns, and experiences that have made up their lives—until this point—created that worldview. Their collective subconscious had been coded differently from that of the generations before them. It also helped that they were in that brief sweet spot of neutrality, the fleeting moment in time between running away from girl cooties and pulling pigtails.

When thinking about our minds, we can imagine them in two parts. The part that moves us to choose a movie, flavor of pizza, or which friend group to hang out with is the more thinking, conscious part. It's driven by a combination of logic, action, assessment, and decisions. It's the part that motivated you to choose to buy this book. It's the awake part.

The other, more significant, part of our mind is the subconscious motivation in our brain. It is what's actually running our lives, affecting our behaviors, beliefs, and habits that are so ingrained that we aren't aware of thinking about them. Everything the subconscious does is about keeping us safe. It does this by attempting to make sense of our experiences, the instructions we are unknowingly giving it. Our subconscious frames how we see the world based on how the world was shown to us. For argument's sake, we will call this the asleep part.

In the boys' world (at this point in their lives)—and this is a tip of the hat to the school and communities Hays is a part of because I know this is not always the case—boys and girls did the same things and did them *together*. Teachers weren't just women, politicians weren't just men, toys weren't divided, colors weren't overt signals, and career paths weren't distinct to a gender—women were doctors and men were nurses. Outside his immediate world, Hays is reaping the benefits of changed structures where better representation of females on boards, in media, and in politics trickles down into everything from the shows he watches to the person he sees doing business with his parents on the other end of a Zoom call. Positioning women in those influential positions has started to make ripples of change from their hierarchy. The influence of having a woman director behind the camera of a movie changes the literal view of how a film is filmed.

Though things have improved, they're far from fixed. Women are still only 27 percent of Congress and as CEOs in top-performing companies, only 15 percent. But to Hays and his friends, it's completely normal, and expected, for a woman to run for president. It's amazing how much even a little representation can start to shift things.

The landscape is not where it needs to be, though our corner of Brooklyn might be inching closer. The everyday uphill battle of raising good kids in our confusing society often has parents feeling exhausted or beaten down. But here, on this movie night, was hope for the future! These boys didn't see girls as separate or less than them. I realized with glee that the influences this generation was under had shifted their subconscious! There was much less gender bias. Whatever it was—sports, parenting, teachers, movies, or the effects of a cultural shift finally taking hold—it was working. The evidence was in these five young men being absolutely gobsmacked that a girl wouldn't be allowed to play with the boys. The future was going to be A-OK, maybe even equal. And I couldn't wait to tell their parents.

A group message filled with celebration emojis and bursting hearts made me feel lighter, happier. With progress came proof that our uphill battles had been worth it! What's a little exhaustion when we can finally break the wheel by not perpetuating the subconscious and unconscious biases of our own generations?

The ways we influence are a part of our subconscious built upon a lifetime of also living under subconscious influences. So we must be careful with them. We need to bring them to the surface, wake up to them, so they are subconscious no more. Because if we don't, our dated, deeply held unconscious beliefs from our own subconscious influences will teach others whether we do so intentionally or not. And in some cases, it's the overwriting of an existing subconscious belief—like girls and boys are equal. With that influence (exerted purposefully or not), we don't create a new *sub*conscious bias, where we overwrote the previous one, but rather a *conscious* one—one that was taught, shared, or wielded to conflict with our previously held beliefs. It's no longer something we are unaware of. It's chosen.

Not long after I sent that celebratory group text, subconscious bias rose to the surface to remind me that, unfortunately, others' subconscious biases teach too.

It was the scoreline heard round the world—13–0 in the now infamous 2019 FIFA Women's World Cup, the United States versus Thailand match. It's a game that's known now more for the fallout, the headlines, and the criticisms than the actual playing, unfortunately.

For our soccer-playing kids, we (moms) organized the viewing ourselves. (Hays's coach didn't plan anything for the match, though we reminded him he had done so for the Men's World Cup. "My bad," he said.)

Not ones to let that blind spot result in our kids' missing this moment, we gathered almost the whole team (now divided by genders at this age) at a local establishment with a big screen and big beers (for us) and sat shoulder to shoulder, moms and sons plus one dad, amid bowls of french fries, soccer balls, and American flags.

The boys were all in. Their pride was evident in their blotchy faces and sweaty foreheads. Possessed by the possibility of winning, of being *the best* (because an eight-year-old loves nothing more than those bragging rights), they compared notes on Megan Rapinoe, Alex Morgan, and Ali Krieger the same way they did Harry Kane, Kylian Mbappe, and Cristiano Ronaldo. *"Who do you think is the best?" "Which one will score the most goals? Is she the fastest one?"* And they'd turn to us, asking, *"Do they have cards to collect?" "How about sticker books?" "Which one is your favorite?" "Can I get a Rapinoe jersey?"* Watching these kids watch these women, we couldn't help but notice that they didn't see *women*, they saw *athletes*. And these athletes weren't just winning; they were making the impossible possible.

The room erupted louder with every goal. The boys danced with us, leaped into each other's arms, screaming, jumping up and down, almost knocking our beers over in their uninhibited excitement. Squeals of delight and shock abounded, one after another and then another. They cheered and chanted, "*USA! USA! USA!*" and when the boys heard that the team could potentially break a world record, they stood inches from the screen, hands covering their mouths. Hays—much like me—gets choked up in life's most unifying, electrifying moments where the solidarity of those around you sharing the same feeling is almost too much to bear. He wasn't the only one close to tears. We moms were too.

Word had quickly spread across social media. We shared text messages of red, white, and blue flags and videos of the kids on the group soccer team chat. Sounds of cheering could be heard in the streets. So as the game went on, some of the dads began trickling in. We thought they wanted to join the fun as they ordered beers, and we scooched over on benches for them to sit. But their body language said otherwise, across the board. At first, I noticed a few crossed arms. Chin up. A deep sigh. A fake smile as their sons looked to them to share in a celebration. An eye roll. And then, when another goal was scored, lifting us all out of our seats once again, they all but one noticeably remained seated.

Why were the men not celebrating with us? Why was their body language closed off, one might say contemptuous, in what appeared to be judgment? I was confused. It couldn't possibly be what it appeared to be because these dads are the very progressives known for championing women's rights, gay rights, diversity, refugees, and so on. These are the men who wear their babies in carriers, change diapers, and don't just babysit their kids. Hell, our neighborhood is the epicenter of woke liberal poster boy dads, the proverbial hub of snowflake central, as the *New York Post* would put it.

And yet here they were, scoffing at women athletes. Not woke at all. Asleep to the fact that their unconscious biases, built on a generation of subconscious gender influences, had them acting like a small-minded coach from a small town in a 1994 kids' movie. They were unconscious of not only their own behaviors but also of the impact of how what they were saying—at first just with their body language, and then growing emboldened by others feeling just like them, with actual words—was coaching our boys. At first, I hoped I was imagining things, but then it became overt. Their conditioning was overwriting their innate instincts. They were unconsciously and, monumentally in this moment, influential. Because not only were they acting out their own influence, but they were also influencing our boys negatively in real time.

Apparently, an ending score of 13–0 was cause for outrage. This time women were accused of being too successful, too self-serving, too athletic. Over the next twenty hours, the words spilled from men— dads, boyfriends, coaches, teachers, the deli guy, the butcher, the bartender, the brother, the commentator, the pundit, the taxicab driver.

They're being unsportsmanlike, they said.

They should consider the other team's feelings, they said.

Think about how it's perceived, they said.

Having such a large point differential reflects badly on women's sports, they said.

Running up a score like that is unnecessary, they said.

They should think about their reputation moving forward, they said.

Why didn't they consider the league growth as a whole? they said.

How unclassy, they said.

So disrespectful of their opponents, they said.

They should've held back and taken their foot off the throttle, they said.

Celebrations are in poor taste, they said.

If they keep doing it, everyone will be against them, they said.

Women's sports need to be held to a different standard to grow, they said.

They didn't consider the unfairness of Thailand having less money, they said.

That score spread makes it disparaging, they said.

Each statement said, shared, megaphoned by pundits, written into headlines, was another drop in the bucket of words women have heard from men their whole lives. The culture had found a way to tear down even the most triumphant women in their most triumphant moment. The success of the female team made people uncomfortable even if they weren't sure why, so they found a way to rationalize their bias. They made it the players' fault. It wasn't until witnessing that now-stolen moment through the eyes of a new generation that I woke up to it.

I've been around sports my whole life as a player, a spectator, and a fan both in actual stadiums and in crowded living rooms and bar-rooms. I have never, *never*, heard these statements before. Not even one. Not in football when the preplanned and practiced end zone dances that happen at every play are vulgar and over-the-top. Not in baseball when the fans start to file out of the stadium in the top of the fifth because it's a bloodbath. Not in basketball when LeBron, Shaq, or Kobe would hang over their opponents' heads from a basketball hoop before running laps literally and figuratively around the competition. Not in tennis when a male icon threw and smashed his racket.

When player Alex Morgan was asked after the game why she kept scoring, allowing her to tie a FIFA Women's World Cup record, her response was: "Every goal counts." It was reported by CNN that "from a sporting perspective, she was right." What other perspective should be taken for a sport? Why—when we should be celebrating the largest victory of the *actual* best team in sports—are reporters talking about feelings and mansplaining how women athletes should behave?

Unconscious bias.

Women are, have been, and—if we don't wake up to it—will continue to be held to different standards that even our most progressive men cannot see. Viewership is up, people watch, and yet the trickle-down effect of influence keeps women's sports off TV or in less than desirable time slots, and without commentary, which doesn't help to build an audience. A USC/Purdue study found that women's sports are almost never featured on TV news and highlights. In 2019, 95 percent of all television coverage, including ESPN's *SportsCenter* highlights show, concentrated on men's sports, according to a survey conducted every five years since 1989. This was true for social media and online sports newsletters. The favorite "argument" of many is that nobody watches women's sports, but if we keep perpetuating the inaccessibility of watching them, then another whole generation won't be introduced to them. To break the cycle of patriarchal influence (*Outer World*) in sports, the girls need to get as level a playing field as the boys.

Sometimes we are more influential than we realize.

The most dangerous ways we influence come from our subconscious, which, as we've discussed, has been influenced. Those unconscious opinions or feelings are built up over a lifetime of living under these subconscious influences, leaving us to influence, override, overwrite—without realizing we are. One important element to remember is that just because we feel uncomfortable with something new, doesn't mean it's inherently bad. If something tests our ingrained beliefs, it's not uncommon to feel defensive, but if we can mitigate that, change will come, for everyone.

My son came home from school the next day, and I told him that we had plans to watch the next USA match with his team again. It landed like a deflated balloon. He made a *humph* sound, turning away from me disinterested. I asked him why he didn't seem excited; didn't he want to watch them again? His reply slayed me.

"Not really, Mom," he said. "Those girls behaved badly."

Those girls behaved badly.

The bad influence can take over, even unintentionally, if we don't wake up. But when we are awake to our own influences and the influence we unconsciously wield, we can help shape the subconscious biases in ourselves and in those who look up to us and then create beautiful moments like antiquated family movie nights where we can discuss outdated *Outer World* structures and how to push back against them. Maybe it'll result in a global World Cup watch party the next time it comes around, one where feelings, whether female or not, take a back seat to the sport itself.

Folklore   You are what you think.

Folktale   I am who I am because of the thoughts in my head.

Folk Rebellion   I realize I am more than what I consciously think and that I have influences deeply ingrained in me that I must wake up to so that I cannot *not* see them anymore. I can look to other communities, generations, information, and educators to expand my understanding of my subconscious self, and in recognizing my own unconscious influences, I can be a better role model (and influence) to those around me.

*Raise Hell   Plan a movie night of inappropriate movies (for today's world), the ones that make you cringe and say, "*You can't do that anymore*." Now, imagine it's ten, twenty, thirty years from now and future generations or your grandchildren are watching our movies of today. Anything in there you think might not age well? It's a fun way to see how things we think are normal might have room for growth.

PICK A FIGHT

"Mom, my friends at school said *divorce* is a bad word."

He shot me the look he's given me his whole life that loosely translates to "*How could you do this to me?*" This sense that I'd somehow failed to give him the same knowledge as his peers has come up before, and since.

There was the Big Bang debacle in fourth grade, when he arrived home from his new school feeling utterly duped and let down that his classmates knew what it was, and he didn't. Why hadn't he been taught about that *like all the other kids*? And at three, when Hays attended a nursery school in a Jewish temple despite not being Jewish (or religious at all, for that matter), he was beside himself to learn that we wouldn't be having a Passover seder. I laughed when I realized what he thought, but it wasn't funny to him. At the crux, I realized too late, his disappointment (by way of a temper tantrum over gefilte fish) was less about the food and more about being different. *What would he say to his friends at school on Monday?*

There is so much to teach when you have a child. There are the things that very obviously fall into the parental bucket: how to eat, talk, walk, and wipe your butt. And then there are the more malleable lessons, things like how to behave in the world: manners, social graces, curiosity, emotional intelligence, and how to build a life. For those, children are subjected to whatever their guardians feel is necessary, not to mention how they behave themselves. This is why there is such a wide range of beliefs and behaviors among kids. There is no universal lesson plan, even for next-door neighbors and classmates. Different zip codes, different houses, different influences all create different structures within the child's home, classrooms, and community.

In our home, I focus my efforts on unlearning. Unlearning is paving the way for the narrative I want to lead, encourage, or embrace. Things I want to course-correct, stand up for, fight for. Around our kitchen table, Hays is unlearning the less helpful things he's inadvertently heard from other adults or other kids, the news, commercials, or osmosed from the times and the culture at large. We're in the habit of teaching him to question the things that are presented as fact when they're actually opinions, mentalities, or accepted norms dressed up and paraded around as indisputable realities. These are the things I have decided to pick fights against.

There are big things like politics, religion, pop culture, music (at least in our house this is a big thing). But then there are the smaller things too, with sometimes larger implications.

> *Practice makes perfect. Life's hard. You're okay. Grow up. You're special. Toughen up.*

I am not one for motivational quotes hanging in our home or on my social media feed. For a cynic like me, they're too falsely uplifting, positive, and feel-good; all I see are empty platitudes posted by people online. Often they're a distant relative of eloquent speeches or poetic prose that have been bastardized by being chopped up and condensed into easily consumable, and easily misunderstood, internet-friendly slogans. The truth is, you may feel that these platitudes and "truth bombs" ring true. Life *is* hard. Everything *feels* stressful. You should suck it up. Get serious. Deal with it. Accept these facts as a part of everyday life. But that's because we live in a time when these thoughts, these phrases, are the accepted norm. And so I feel it's my duty to teach my kids to unlearn these supposed hard facts of life. I refuse to allow these impressions to influence them into believing that giving up and giving in is part of being human. The day my kid comes home saying, "it is what it is," is the day I set fire to it all. But there are a few sayings I *am* known for espousing in our home:

> Easy peasy lemon squeezy. Practice makes progress. It don't matter. It'll be okay—someday. Ain't no thing, chicken wing. Worry is wasted. Piece of cake. The juice is worth the squeeze.

Sure, my words won't be quoted by some scholar in the future, but they singsong in the heads of my children, who I'm desperately trying to counterinfluence. Each of these sayings is a way for me to counteract the influences of the world outside my front door by combating ideas of perfection, anxiety as fuel for success, or that giving up is better than working hard for change. And it's working. Hays most recently has taken to saying, "No worries," when his baby brother, Beau, spills his milk or telling him, "The juice is worth the squeeze" while he screams his head off in the back seat of a long car ride. Beau is still a baby and has no clue what his idol and older brother is saying. But I know he will grow up in a home where messages that breed future unhappy and potentially toxic behaviors will be replaced by positive, less stressful (or empty) affirmations, conveniently relayed by someone besides myself.

Narratives drive the stories of our lives. They're how we experience our reality and remember our memories. And while the narratives outside our walls may not be ours to control, we have a chance to change them, from the inside out, by picking a fight. We have the option of teaching critical thinking so that these "norms" aren't blindly accepted as such.

I wanted Hays's life, reality, and memories of his parents' divorce to be less tragic and pernicious than, say, those of my childhood. I've been attempting to let go of my narrative around that time of my life since I was his age. That's why when Hays came home with his peers pushing back on the fight I chose to pick, I knew how loaded it was.

I won't bore you with the details of my parents' divorce, but my siblings and I are still uncovering the results. I've done lots of work to understand how this influences my *today* as a deep-down people pleaser, a fixer, someone hard to catch. My attachments, my love languages, my (un)comfortability with money, my autonomy, my relationships with my siblings as adults, all ladder back up in some way or another to the divorce. We all have issues to work through, no matter whether our parents ever married, stayed married, or not. Thankfully, mine chose to do this difficult thing instead of staying unhappy and together, which would have laid the groundwork for future misaligned views on what love and partnership should be. My parents did the best they could under the influences of the social norms of the times, their parents, and the upbringings they brought into the union. As an adult, now I can look back, as can they, and more easily see how things went awry and how it probably could've been handled differently. But like zeitgeists, it's hard to realize the things influencing us while we are in it.

And boy, with Hays, was I about to be in it.

When Hays came home talking about how divorce was a unilateral bad thing, I started by writing down the things I did not want Hays to think, experience, or remember. I wrote through the lens of the D-word narratives I'd been living under most of my life. The list ended up being longer than I thought it would be.

I did not want him to think it was his fault, that it made us love him less, or that we are any less of a family than nuclear families are, or that he is any different from any other kid. I did not want him packing and schlepping bags, having one bedroom with a mother's touch and one that felt like it was slapped together by a boy in college, a second home that took him away from friends or felt like a far drive from his community. I didn't want him to have two separate birthday parties, to have to play telephone between two nonspeaking adults. I didn't want him to lose traditions like seeing the Christmas lights or first-day walks to school. I did not want him to feel awkward when speaking to any of his grandparents or feel cut off from them when with the *other* side of the family. I did not want him to hear or know the reasons for our ending. I did not want him to think his father was a bad person or that I thought he was a total shit. I did not want him to feel guilt or shame. I did not want him to play the role of the fixer, or keep secrets, or lie. I did not want him hearing us argue when together or grieving when apart. I did not want his friends to pity him or teachers to stumble over their words when working on family projects. I did not want him to have to say, "I don't know," when asked if he could play because he didn't know which house or which rules or which schedule he would be following that day.

When his father and I would (hopefully) rise from our ashes, introducing a (hopefully) better fit and happier partnership into the mix, I did not want Hays to think this meant he was being replaced, that a new person would become another mom or dad. I wanted him to know there was only one of each of those, but there was always room to grow our modern family with bonus people who would (hopefully) only make our lives richer. And I wanted him to know that if a bonus person wasn't so good, his parents would always have his back, were always a team, and would always put what was best for him first. And I wanted to feel secure that as he grows up, he won't believe all marriages end in divorce, that love isn't real, or that partnerships are best for dancing. I didn't want my divorce to plant the seed of chipped-shoulder independence, setting him on a path where he's most comfortable alone, with stoicism leaving no room for feelings or other people. I didn't want him to be a lone wolf, an island, an uncatchable nomad.

Getting divorced is hard enough. It's impossible to describe the excruciating feeling of unsnarling the lives of two people while maintaining the life of another. Repositioning how we do it, how we talk about it, and how we live it for the benefit of the one who shares our DNA (but hopefully none of the trauma) is progressively more possible though still astonishingly difficult.

Rebranding divorce is an uphill climb when our society has been glugging it as a social tragedy through the influences of culture. Shows like *Divorce Court, Jerry Springer, Maury,* and daytime soaps feed off drama, hate, and infidelity, positioning divorce as a failure of the poor, the uneducated, or the shameless. Alec Baldwin and Kim Bassinger's infamous leaked voice mail and custody battle over their daughter became public domain regardless of a child being at the center of it. The affair and royal divorce of Princess Diana and Prince Charles resulted in paparazzi chasing her to her death and then became watercooler fodder as the spectacle of their lives falling apart sold headlines in the tabloids. And then there are the gender biases: Whether a woman is the leaver or the leavee, she's branded as either an eccentric slut, a bitter unloving old battle axe, or an ungrateful bitch. The D-word is one that comes with a loaded bag of assumptions filled with hate, drama, gossip, and judgment and none of the actual context of what transpired behind closed doors.

Having kids with someone who you are no longer living with, married to, or romantically involved with keeps you tied in a relationship no matter what—for life. Whether you like them or not (or that fact) won't change it. So, how that inescapable relationship unfolds is usually up to the divorced parents. And how we presented ours to the communities all three of us were involved in was up to us. After all, our child would feel the effects of our behavior for his entire life. And so, once again, I swallowed it all and straightened my shoulders to shoulder the burden for us all.

I was so hell-bent on not repeating history, I *finally* took Gwyneth Paltrow's advice. I decided to rebrand our divorce, à la *goop*'s announcement of her and Chris Martin's now-infamous "conscious uncoupling," thus influencing how my son experienced it.

Stepping into my influential power—this time for good—I stood up for what I thought was right. I rewrote the narrative.

I wanted Hays to have a room that felt like his, in a neighborhood that was his. So we chose to live within walking distance and spent time decorating and stocking the rooms to resemble each other. No schlepping. Our relationship is one of proximity, but it is more than that. We chose to tell Hays that our relationship was still a *family* filled with love—it just looked different from what others might expect. When his friends ask about us, and they all do, he says, "My mom and dad love each other but can't live together," not being ashamed of the D-word. This can be confusing to outsiders, especially to children who operate in black and white, but it was not confusing to us.

As Hays moved through grades and his peers became older, wiser, more influenced by their surroundings and society, they started to doubt him and his assuredness about whether his parents were really friends. In truth, around the same time, so did I. He came home after a dreaded school assignment in third grade about families: drawing his home—singular. He asked his teacher for help, letting her know that his home was both places and asking how he should draw it. They got it handled, but it perked the ears of fellow classmates.

"Oh, your parents hate each other?" one kid asked. *"Your mom and dad must fight all the time!"* said another. *"Divorce is a bad word"* was the final blow. We don't say bad words in our home. Either of them.

I had to try to explain to a crying eight-year-old, when I'd already lost his trust, that while *other* people think that it is bad, we don't. I tried to tell him that for us it was good. It let him grow up without the problems of his parents to mimic when he was older. It saved us from using words, tones, and body language that said one thing but meant another. It gave us new relationships in which he could see what partnership should actually be. None of my thoughtful explanations mattered in the wake of one of Hays's first big experiences with pushback, though. I left him open to ridicule and scorn—the same as when I was young—and he felt duped. Only this time it wasn't gefilte fish. It was his identity, his home, his family, his life.

What started in our home couldn't always stay there. When you pick a fight to rewrite a narrative so embedded in our collective consciousnesses, the trickle-down influence takes on a life of its own. It bled into his world outside our four walls, and the influence of outside norms bled into our little corner too. I could control the narrative among us, but once out in the real world, it would answer back. That didn't mean we couldn't continue to stand our ground until the *Outer World* caught up and continue fighting for the narrative that was true to us (our *Inner World*), though. We could counterinfluence our *Surface World* too.

Hays's move to a new school during the pandemic was just another notch in the long line of changes influencing who he would become. Ultimately, it was a positive influence in line with our rebranded narrative, and it showed him how to voice, and be proud of, who he was, and where he was from. Two houses, one home.

Where I Am From
by Hays

I am from suburban greens
Lightning bugs, shooting stars
I am from busy city streets
Birds chirping, cars honking.

I am from "on the count of three!"
And "easy peasy lemon squeezy"
I am from Papi, Tata, Nana
Meema, Grumpa, BunBun.

I am from two houses
Mama's rule breaking
Messy Jessy fun, Press Your Luck
Big couch cozy cuddles.

Dada's rule following
Tidy and perfect, Sky-Jo
Roughhousing "you knucklehead!"

I am from "Utica Rules"
Monopoly, Chess, and cards
Camping, dancing and listening to Bowie.

I am from Red Hook sea glass
Crystals and rocks.
Friday pizza movie nights and
Ice-cream sundaes on the first day of school.

I am from "piece of cake!" and
I love you more than the moon and the stars.
Lovebug, Bubba, Pumpkin, Kookachoo.

I am from valuing everyone's opinions, trust,
no lying, enjoyment and happiness.
I am from too big,
Big dogs, big feet, big hearts.

I am from dog-eared Dog Man
Winning and being undefeated
Tottenham soccer club, Bunk'd,
Wimpy Kid, and Saturday morning cartoons.

I am from clocks ticking, subways screeching,
Sun shining, dogs barking, people walking,
Strangers talking, kids running,
Leaves blowing.
I am from Brooklyn.

Big Bang theory or not, I think we made the right choice—both in switching schools and in rebranding divorce for our fugazi family. But I also made an intentional choice because I didn't want Hays to think marriage is inherently dysfunctional; I didn't want our relationship to be his model for love and respect. I never thought past that goal. When you're in it like that, you can't see past surviving the day. But by switching what I wanted from a partnership, I was able to finally give my son something I could never have given him on my own. When my now-husband, Mikey, started to spend more time around me and Hays, the impact of seeing two people in love, who speak kindly to each other, laugh often, and are true friends was unexpectedly profound. It influenced his perspective right away. We were riding in the car one day, picking on each other in that way you do when you are all settled and content and happy, and Hays said from the back seat, "If I get married, I want to match—like you two." Now, that was the kind of influence I wanted to have on my son.

Today, years after the fall, I've become a pseudo welcome wagon, the first stop for women who are looking to leave their marriages but not leave a wake of divorce (in a traditional sense) behind them.

My influence on Hays has had a larger impact in our small community too. Teachers, partners, other parents, and our own divorced parents have had to rejigger how they think of divorce; they view our family now through our reality, and by default, they've reassessed their own influences. In terms of adjusting to and accepting our rebranding, I'd say it's been 100 percent smooth sailing, but that would mean that I am 100 percent full of shit. Some days I'd love to do or say or be all the horrible things that are hallmarks of traditional divorce. It would feel great to let that fuck-it flag fly. But then I stop and think of the implications of that in my endgame fight. Truthfully, I've had to rewrite this essay to be more evergreen and not write from the constant swinging pendulum of emotions, always under the influence of other people and the influences (*Inner, Surface,* and *Outer*) they are under within my little experiment. But for now Hays has mostly survived the breakup and the rebrand. Thriving? Only time will tell. He, too, has friends whose parents are going through it "like we did, Mama." He no longer feels alone or different; in fact, he's offered a bit of positive influence in his community as well, and when the narrative shifting fails for his friends going through it, he turns to traditional D-word platitudes. *Double presents. Two bedrooms.* Some things never change, but lots can. We can influence not just our small communities but small children, and their communities too.

Folklore You must be a "person of influence" to be influential.

Folktale Without fame, money, power, status, I have no influence.

Folk Rebellion I should not underestimate myself but realize the responsibility and opportunity to take a stand with the massive influence I hold within my own social circles, personal communities (work, school, hobbies networks), and family.

*Raise Hell Get out a piece of paper and write "Mad as hell" at the top. Write a list of things that make you mad as hell. Can be anything. Now look at that list and see if there's anything on it you want to pick a fight over. Start with small changes that can be made in your immediate circles. You will begin to see you can create a ripple effect of good influence outside of yourself.

BAND TOGETHER

To this day, my four siblings and I are often referred to in unison, as "*the Elefantes*," as if we are one living, breathing organism instead of four separate ones. Large families in small towns end up with a sort of local notoriety by default. I was the oldest. Growing up, my younger siblings complained of having a reputation that preceded them. I didn't think it was so bad, but maybe that's because as the oldest I never had to experience an incredulous teacher getting to the Es during roll call on the first day of school remarking, "Another one?!" Being lumped together, not only by DNA but also by assumptions, can sometimes have less than desirable outcomes. But mostly we found that being a part of a group (or a gang, as some referred to us) was incredibly beneficial and, most of all, fun.

We learned a lesson very early on that still rings true into adulthood—when you band together, you can force the hands of those around you more easily than if you were on your own. Having more voices and more allies meant that we were more likely to get our way. The bigger the ask, the more of us were needed to shift the tide. Some things required the full gang to have one another's back. So when we discovered that Woodstock '99 was coming to our little ole town in Upstate New York, our parents never had a chance.

Strength in numbers is what allowed me to be among one of the hundreds of thousands of people in attendance at the ill-fated festival.

I also had my brother in tow. Strength in numbers is also how the historic concert series went from a peace-and-love fest to an angry and disillusioned frat fest in a weekend. Whether it be four siblings or an audience half a million strong, when people come together in groups, acting synergistically can create a whatever-the-hive-wants mentality. In science, this is an actual phenomenon known as a "superorganism." It refers to a social unit of species like ants, bees, or termites in which the collective acts in concert. Often due to challenging environments, the group finds that its individuals are not able to survive by themselves. So the group creates a division of labor—like choosing a new nesting site or finding and creating new nutrients—in order to live. The term was an apt description for us concertgoers that weekend. In our case, what we needed at the festival was water.

Survival is a pretty darn good reason to collectively wield influence, if you ask me, and according to perhaps the most well-known theory of motivation, Maslow's hierarchy of needs, our most basic needs are fundamental to survival. In Maslow's hierarchy, psychologist Abraham Maslow showed that people are driven to take care of these particular needs before being able to move on to their more complex ones and reach their full potential (the ultimate goal).

There are five different levels in Maslow's hierarchy, starting at the lowest level, the base of the triangle, with physiological needs that are vital to survival. Kind of like water at a festival on a tarmac in 100-degree temperatures. Some other examples of physiological needs include food, air, shelter, sleep, clothing, and reproduction (because they're essential to propagating the species).

At the next level up, the needs of safety and security come into play, which can be more nuanced because these elements are defined differently by different people. Maslow found that financial security, employment, resources, personal bodily safety, health, and property lead to feelings of control and order. Shifts such as moving to a safer place to live, saving money, getting a new job with better health insurance are all motivated by tier two. These two bottom levels together are referred to as basic needs for all humans.

Levels three, four, and five are social (love, acceptance, intimacy, and belonging through friendship, family, partnership), esteem (recognition, status, respect, strength, freedom through hobbies, accomplishments, and purpose), and self-actualization—becoming the most that one can be. According to Maslow, reaching the pinnacle (the tippy top of the triangle) of self-actualization "may be loosely described as the full use and exploitation of talents, capabilities, potentialities, etc. Such people seem to be fulfilling themselves and to be doing the best that they are capable of doing. They are people who have developed or are developing to the full stature of which they are capable." At this final level, a person has moved beyond the need for accolades, accomplishment, and prestige to a whole new level of being fully self-aware, less concerned with the opinions of others, and motivated by self-fulfillment.

Feeling stuck and unmotivated might be a symptom of being unable to progress through Maslow's hierarchy. And at some point, we will all get stuck or toggle between the levels due to divorce, job loss, change of circumstances, etc., because life is not linear. But when a person, a group, or society's most critical needs aren't being met, there is a deficiency. This deprivation, if not addressed, will lead to consequences like collective unhappiness, but also illness, particularly rampant psychiatric illness or mental health issues, death, or other trauma.

Do I condone the mob mentality at Woodstock '99? Absolutely not—but maybe we can learn from it.

The tension had been rising with the temperatures all weekend, and there was a gnawing in the pit of my stomach I could no longer ignore. Everything in my body told me to find my people and go. My reptilian instinct to flee superseded my desire to see the Red Hot Chili Peppers. As we wove our way out to a raucous rendition of Jimi Hendrix's "Fire," officers in riot gear were lining up to come in. Things had escalated quickly, and the overflowing, dehydrated—already hyped—crowds had gone wild. Fires broke out, cars got flipped, and vendor booths and tent cities were destroyed in an act of mob fury. Maybe we didn't see it coming, even though the signs were all there, but you could feel it building.

For me, it's kind of hard not to make a comparison with our current times. I've got that feeling in my gut again. Rising tensions. Angry mobs. Maybe if we looked to Maslow's hierarchy of needs, we could make the connection in moments when our society is starting to feel more like a trainwreck festival than a functional superorganism. Too many people don't have what they need; too many people feel left behind. If affordable healthcare, financial security, bodily safety, and autonomy are not on offer, though we've been told they're our inalienable rights (or Maslow's basic needs), the end may just be a society that feels an awful lot like the crowd at Woodstock '99.

It was horrifying to see a tractor trailer tipped over onto a person when I was leaving the festival in 1999. And I was equally stunned on January 6, 2020, when an angry mob stormed the United States Capitol under the influence of a persuasive and misleading leader as well as the pull of the hive-like superorganism mentality. Both incidents of violence demonstrated what can happen when people come together under duress, driven by anger and fear, to get what they want or have their needs met.

We're clearly at a tipping point.

But there are other more positive alternatives to channel the hive mind in constructive ways to try to get us all what we need. And, let me tell you, it's definitely not about someone else taking what's yours. If we band together, and show compassion and concern, there should be enough for everyone. And in the case that there isn't, then it's an implication of a much larger issue—a broken or unjust system—and then our only chance *is* to come together to correct it.

Invoking the power of collective influence to force the hands of entities or structures that have negatively impacted our lives for decades is incredibly potent. Groups of angry, fed up, motivated people have changed the course of history time and time again by organizing. Collectively opting out of something, boycotting, setting up picket lines, staging sit-ins, and marching are all examples of guerrilla activism. "Power to the people" is an expression that's been used around the globe—because it works. In the 1960s, young Americans began using this expression as a sort of protest against the older generation, notably "the establishment." Black Panthers used "All Power to the People" to challenge ruling class dominance. Students protested the Vietnam War with it too.

Radical change is possible when people come together. It's powerful in tipping power.

Advocacy, organization, and activism allow one voice to become many. The hive-mind superorganism sets about on a shared mission to change anything from laws and norms to perceptions and infrastructures. Today "Power to the People" doesn't just mean gathering in real life to march, knock on doors, petition, or canvass. With the invention of the internet, it can look like an open-sourced spreadsheet, a viral hashtag with a call to action, or a safe place for marginalized groups to have a voice.

Part of what keeps people under the influence of the frameworks of previous generations is that they exist in the structures set forth by them. There are unspoken, sometimes even spoken and mandated, rules to which people are expected to adhere. When someone says, "That's just the way it is," or that a certain "rule" is not to be broken, the "power *of* the people" not only questions who set the rule and why, but also lifts the opaque veil of secrecy.

Pay disparity is a modern-day example of how we can band together in action to force out those old patriarchal systems. Money. Earning. Salaries. Our culture has ingrained in us that we are not to talk about money, especially when it comes to what we earn, which leaves workers at a disadvantage. The companies can hide behind their larger identity as businesses, but individuals are not supposed to be open about salaries. This taboo has kept power in the hands of those doing the hiring and continues the structures of elevating white men through higher pay. This norm is an example of a structural influence—meaning that when we perpetuate the structure, it trickles down into every aspect of our lives and society as a whole. By lifting one group above others, we are pushing down everyone else, which means the wheel of basic needs is not being met for specific people.

Activism on the internet has been a huge positive to come out of the digital age. The anonymous sharing of what was once considered a no-no has taken over the internet ever since 2015, when an engineer at Google used her own employer's platform to reveal pay inequities at the company. And so began "Google-doc" activism, which showed that even at Google, a tech behemoth looking to advance and influence society for the greater good, women are paid less than their male counterparts and are underrepresented in influential leadership roles. Despite, or some could argue because of, the continued rise in wealth and power, major corporations are still operating with wide gender disparities, as proven by the McKinsey Global Institute analysis conducted globally.

Today crowdsourced information by way of an anonymous spreadsheet has been used not just for pay in a range of industries, but as with all things web based, it's also being co-opted and expanded. Now Google-doc activism has been used for countless pursuits. One titled "Shitty Media Men" was created in the wake of the #MeToo movement as an online safe space to name men accused of everything from skeezy flirting to rape; it has since been taken down and lawsuits are ongoing. There are lists for helping people in need after disaster strikes, lists for police violence and misconduct across America, and lists for how to survive provided by local mutual aids during pandemics, presidencies, or recessions. By being simple to use and free from gatekeeping, Google docs has become an online action hub for people who want to band together to enact change.

Facebook originally set out to "give people the power to share and make the world more open and connected," but their mission had one fundamental flaw—it didn't expect or encourage any specific positive outcomes from connection. In 2017, it pivoted with this in mind to "give people the power to build community and bring the world closer together." Today controlling the narrative for the metaverse is generally shortened to "bring the world closer together," leaving community out of it completely. But community is what the internet has always been great at doing—banding together groups of people and providing them with a hub to speak freely, a sense of safety, and a place to belong.

What we do with those crowdsourced spaces of togetherness is very similar to the crowds at festivals. The "vibes" shift based on the needs of the collective. Sometimes it's wielding influence for basic needs like advocacy for climate change so that our species may continue living, and other times it's wielding online influence just because we can. Regardless, our needs for togetherness can be as strong as water or food.

When Foo Fighters superfan Fabio Zaffagnini wanted to see his favorite band play for his tiny village in Italy's countryside, he used the power of people *and* the power of the internet, acting in concert, to get his concert. In a rather extraordinary use of the World Wide Web, he spent one year crowdsourcing and crowdfunding fellow Foo fans, musicians, and an audience to cheer his mission on. His idea wasn't so simple, but if he could pull it off, the megaphone would hopefully be too hard to ignore. He decided to create the Rockin' 1000—one thousand guitarists, drummers, and singers—to play the Foo Fighters' song "Learn to Fly." Simultaneously.

It took one year of planning, $50,000, and a bevy of allies, all of whom were crowdsourced through the internet, resulting in a now-viral video of one thousand people in a field playing the song as a tribute—but also an ask. With more than sixty million views to date, it was impossible for the Foo not to see it, and not be inspired by it. The band said they wanted to play Cesena, Italy, and they were as good as their word. They performed for the town, opening with "Learning to Fly" with lead singer Dave Grohl admitting he cried when he saw the Rockin' 1000's rendition. Fabio's plan worked; the dream had been met. But besides its purpose, the ripple effect of its influence lives on in the community it created in real life. The Rockin' 1000, the Biggest Rock Band on Earth, now plays all over the world. Rivaroli has said of its unexpected success: "There is a lot of ego in music, but this project tried to take it out of the equation. Everyone had to be on the same level for it to work."

As individuals, we can band together as pseudo superorganisms to create a level playing field, enacting a type of influence that continues to live on in perpetuity. And we can organize to stand against all things that prevent us from having our basic needs being met. Whether it's advocating for justice and reform in systems and cultural narratives that divide and oppress, or a simple want—togetherness and some fucking rock 'n' roll—strength in numbers, in person or in the digital realm, is pretty hard to ignore.

Sometimes, actions do speak louder than words.

Folklore You can only rely on yourself.

Folktale The world is a competitive place. I need to look out for me.

Folk Rebellion I now see that there is strength in numbers, and I don't have to go it alone for the needs I have or changes I want. Many hands make lighter work—and wield a bigger or faster influence.

*Raise Hell This is a time when the internet can be used for good! What sort of things did the transparency or the connecting of the internet help you with in your past? Clarity on salaries? Ability to get signatures for something you were going door-to-door for? If you've got an axe to grind, can you use the internet to find others with the same mission?

MAKE TROUBLE

I could not think of a role model when I was tasked with the homework assignment "Write about someone who inspires you." The people who inspired me were the people who railed against a system that tried to tell them who to be and how to live. My heroes were somewhat antiheroes—disobedient eccentrics, weirdo artists, transgressive authors who loathed authority and felt confined by social norms.

Even at a young age, I was all too aware that turning in a paper about John Waters, the creator of cult favorites *Cry-Baby* and *Hairspray,* which at the time were banned but today are praised, or Charles Bukowski, who famously said, "We are here to unlearn the teachings of the church, state, and our education system," was sure not only to get me a low grade but also have me flagged as a lowlife. So I picked Theodor Geisel, thinking it was a great middle of the road for my thesis—acceptability and firebrand.

He was a curmudgeonly idealist who devoted most of his creative life to opening people's eyes to a broken society through his illustrations, stories, and books. Yup, beloved children's author Dr. Seuss. To this day, many people don't realize he was advocating for social change through parables about the environment in *The Lorax* or the arms race in *The Butter Battle Book.* To me, his brilliance was in how he used his immense talent to share his deeply held convictions about things like fairness and acceptance and influence people without their realizing it—making him one of the most influential people of our times. I thought it was a great angle.

I worked very hard on that smart and well-written paper only to receive it back with "*Dr. Seuss??*" scrawled in the top corner. A sad face after the double question marks was followed by a low grade not reflective of anything except my teacher's disapproval of such an unserious role model. The brownnoser next to me flashed me the A plus on their postured choice—Mother Teresa. No offense to Mother Teresa. She was incredibly influential, but the choice was painfully obvious.

Sometimes the person who can enact the most change isn't always the most obvious choice. Sometimes they're downright sinners, not saints. Or at least they start out that way in the eyes of others, because to break a mold, defy a norm, or restructure a structure, you need to ruffle a few feathers along the way. The most influential people in history were often considered mad because of their grand ideas. Some got their due while they were alive to enjoy it, like when an invention such as the lightbulb finally worked after a thousand failed attempts. Others didn't live to see their genius go from scorned to celebrated. It takes humans a while to come around to new ideas. To do good things in the long run means you might be considered an outlier in the present moment.

By nature, rebels are almost always underdogs fighting back against something they see as oppressive. As their influence shifts a collective consciousness, they end up gaining some sort of respectability in retrospect. While these future role models may not be well thought of in the moment—as they're breaking rules, picking fights, and standing up—it is their inherent selflessness that made them good all along. Influential people are often troublemakers who felt stifled by the system as it was. Throughout history, the artists, innovators, pioneers, intellectuals, leaders, and icons who show up with their bad selves as agent provocateurs blaze their own way, and by doing so, open pathways for others.

Conformity is a word that refers to the propensity of a person to follow the unwritten norms or behaviors of the social group to which they belong. For a very long time, academics and psychologists have been interested in determining the degree to which individuals conform to societal standards or rebel against them.

Troublemakers are labeled as such because they speak up when others won't. Troublemakers are alternate voices. Troublemakers stand up against the people, the structures, and the that's-just-the-way-things-are mentality when nobody else will. They don't conform—and by doing so, or not doing so, become influential.

Psychologist Solomon Asch performed a series of tests during the 1950s known as the Asch conformity experiments, where he discovered that (many!) individuals would rather lie or offer an inaccurate response than stand out. It's the troublemakers who save us from ourselves, the very ones often weeded out of important influential places like government offices and executive boardrooms. When the majority remain silent, even knowing that there is something wrong, it's the handful of noisy, pesky, loud-mouthed dissenters who keep us honest. And those voices, those people, are not always welcome. What is a hero to some is a complete pain in the ass to others. Usually those with something (money) to lose.

Today it's hard to picture driving a car without seatbelts or airbags but that was very much the norm before a troublemaking car executive decided to sound the horn on an industry that was putting sales above its consumers' safety. Aptly nicknamed Nader the Crusader, Ralph Nader is responsible for the automobile safety manufacturing laws we see today, including seatbelts. In a groundbreaking investigation of the auto industry, he was able to show how car manufacturers were putting their drivers in danger. His book showcasing his findings had a massive effect on the auto industry and society as a whole, resulting in the creation of the U.S. Department of Transportation and millions of saved lives! Not surprisingly, some automakers, especially General Motors (then the most powerful corporation in the world), attempted to downplay Nader's critiques, going so far as to hire private investigators to dig up dirt on him in the hopes of creating roadblocks for his mission by smearing his name, but the "crusader" persisted and eventually persevered. New crashworthy standards around brakes, tires, seatbelts, and airbags were proposed and later adopted universally. In 1967, Henry Ford II warned that the standards "would shut down the industry"—because, of course, in America we think first of saving industries before saving people. It took ten years, but Ford finally conceded, noting on NBC's *Meet the Press,* "We wouldn't have the kinds of safety built into automobiles that we have had unless there had been a federal law." Alas, nothing to fear here! The auto industry that wouldn't act in good faith without the regulation that forced them to is still alive and kickin' as are millions of people, thanks to the good influence of a troublemaker motivated by selflessness.

Decades later Nader's antiestablishment mentality and for-the-people advocacy would bring him into the world of politics, making him a key and controversial figure in the 2000 U.S. presidential election—tarnishing his reputation in the process. But the 2006 film *An Unreasonable Man* set out to restore his reputation and legacy. The title took its name from the playwright, critic, and political activist George Bernard Shaw's quote: "*The reasonable man adapts himself to the world; the unreasonable man persists in trying to adapt the world to himself. Therefore, all progress depends on the unreasonable man.*" Shaw's notion that it takes the "unreasonable man"—the one who refuses to accept the world as it is—to influence future progress is not just profound but logical.

During my speaking engagements for *Folk Rebellion,* when I likened the "safety" of the internet to that of cars before serious regulations, people would say I was being unreasonable. *Unreasonable,* even though we are metaphorically drunk at the wheel, kids rolling around willy-nilly in the back seat, drivers running through intersections at high speeds with no cares in the world. The tech industry is like the auto industry of the 1960s. They don't want any troublemakers to bring regulation into their free-market windfall. This time we might not be driving headfirst into one another, but I think everyone can agree we are in a collective crash and burn. Fortunately, there are troublemakers beginning to step up in this area too.

Whistleblowers are some of the most important, charitable, and good people of our time. They sacrifice themselves to hold accountable those with too much power who use it illicitly at the cost of others. This time, saving us from ourselves is an ex–Facebook manager, data engineer, scientist turned whistleblower, and child advocate named Frances Haugen. Originally joining Facebook with an interest in misinformation after someone close to her became radicalized by the platform, she said she "felt compelled to take an active role in creating a better, less toxic, Facebook." What she saw once on board was a pattern of prioritizing profit over public safety, and her attempts to make changes from the inside as a product manager in the "civic integrity department" were met with stonewalling. On October 5, 2021, Haugen testified before the Subcommittee on Consumer Protection, Product Safety, and Data Security of the United States Senate's Commerce Committee. Not so ironically, this is one of the many committees that were created in part thanks to the "father of the consumer protection movement." Yup. Ralph Nader.

Haugen's legacy, her influence on our world, is still being written, but her focus on enhancing awareness of the harms of social media and big tech through government regulations and litigation strategies has already created a tidal wave of change. In 2021, she appeared before the European Parliament's Committee on Consumer Protection. In the hearing, she urged the parliament, which was debating the Digital Services Act—a legislative proposal by the European Commission regarding illegal content, disinformation, and transparent advertising—to mandate that social media platforms operate transparently and not create loopholes that big tech could exploit. She said they had "the potential to be a global gold standard" and an inspiration for other countries on safeguarding democracy on social media. In 2022, an agreement was made. Her disclosure of tens of thousands of Facebook's internal documents to the Securities and Exchange Commission and *The Wall Street Journal* started a rebellion against harmful things influencing everyday people in their everyday lives. For her selflessness in her advocacy for children, she has been honored with the Fred Rogers Integrity Award, named after yet another countercultural troublemaker, Mister Rogers.

Don't let the cardigan sweater fool you. His soft voice and gentle puppets were a means to an end. He hated TV. When he turned it on, he was disgusted by the things he saw—violence and people demeaning one another. To counter what he felt was televised inhumanity, he decided to make change within the medium. Was Mr. Rogers bad? No, absolutely not, but the decisive way he chose to counterinfluence the state of programming by operating as a rogue rebel within it is pretty badass. In the same way Dr. Seuss used cartoons to soften (some would say hide) the delivery of his perspectives on provocative topics, Fred Rogers used his platform to bring about conversations around challenging subjects like war, gender, race, and poverty. Instead of bringing fear into our homes by way of a child's face on a milk carton, *Mister Rogers' Neighborhood* broadcast patience, love, and understanding into the living rooms of young children—and their parents—during a time rife with uncertainty. He and his hokey, low-budget program, with its gentle advice for children, is continually revisited because of its lasting impact. He's been awarded the Presidential Medal of Freedom and a Lifetime Achievement Emmy in addition to honorary degrees and other awards for his contributions. As a visionary leader for children's well-being, the effects of his work to influence television programming are still being understood.

To become influential far beyond your individual self is to create a legacy of influence in your wake. It can benefit a few to many, and in the most influential cases, it can affect not just many people, but many generations and formats, making everlasting changes.

With so many ways to gain influence and wield it these days, people have the power to shape communities, shape culture, and shape the world. Hopefully, they'll choose to use it for good. Becoming truly influential isn't seeking fame, fortune, or an A plus on a paper. It's being considered unreasonable—and standing up in the face of the status quo.

When John Waters, once rebuffed and then regaled for his boundary-pushing films, delivered his advice to the young graduates of the Rhode Island School of Design in a commencement address he'd been invited to give, it was so daring in its mission that it immediately went viral:

"*Hairspray* is the only really devious movie I ever made. The musical based on it is now being performed in practically every high school in America—and nobody seems to notice it's a show with two men singing a love song to each other that also encourages white teen girls to date black guys. *Hairspray* is a Trojan horse: it snuck into Middle America and never got caught. You can do the same thing."

Being a "good influence" isn't really objective and based on each person's idea of good; if it were, we'd be here all day. I am very sure there are people who still feel John Waters is a bad influence. But when *good* is defined by a selfless motivator looking to help others and not just their own self-interest, it's easier to recognize, harder to argue against, and a clear model for us to follow as we strive to be a good influence in the world.

Waters's speech, titled *Make Trouble,* was later turned into a book. In it he encourages people to outrage critics, be noisy, and embrace chaos no matter what they choose to do with their lives.

Or in Waters's own words, "It's time to get busy. It's your turn to cause trouble but this time in the real world, and this time from the inside. Go out in the world and fuck it up beautifully."

Folklore   Be good.

Folktale   I must follow the rules.

Folk Rebellion   Being on the side of good, as defined by constructs/structures/norms, can mean following the rules, but being a selflessly motivated troublemaker and challenging those rules might be the truly good thing to do.

*Raise Hell   What "rules" do you follow that might be roadblocks to making good trouble? Is there something to break you out of the accepted norms that would help in ruffling feathers? Think of the fight you chose to pick; what could you do to be the much-needed troublemaker?

Pay Attention
Mother Fuckers

—Bruce Nauman

Part VI
Above the Influence

Welcome to the final (phew!) part. You made it!

First, I want to thank you for trusting in me to guide you here. As a former bullshit artist who's benefited from influence, I wasn't so sure you would accept my offering and apology. I appreciate your giving me the opportunity to be an inspiration, or good influence, for you. The responsibility is immense, and I take great pride in our journey together.

My hope in revealing the ways influence perpetuates cycles is that it no longer makes it difficult to distinguish or opt out of. Perhaps these revelations are like a vaccine to inoculate you against some of the external influences in your life. Are you immune? No. Resisting all attempts of influence wouldn't be the goal! Persuasion and influence are at the core of a functioning society, and so some persuasive attempts must get through. It's more like you are deprogrammed and now stronger because of this treatment. You can now choose to protect yourself from appeals, demands, and propositions that you feel do not have your best interests at heart, or you believe to be dishonest, ethically iffy. And you can now determine if they are any of those things because you have learned how to do so!

In the following pages, you will be taken through a how-to guide derived from my own experiences and lessons and based on the foundations we have explored up to this point. This section is a workbook about embracing self-determination and independence, learning to dissect the actions of others to get to the source of their motivations, and using critical thinking to live an intentional life. Each essay title will tell you exactly what tactic we will be covering—a technique, a mindset, or an exercise. While we've explored tips and tricks for recognizing and rejecting bad influences throughout this book, what follows is your ultimate go-to tool kit for a life well lived (however you define that), free from the influences that keep you under. You will notice the structural change in the following essays: There are no more *Raise Hell prompts. That is because the whole essay is a prompt now. Breaking the rules, going against the grain, bucking the status quo—

call it whatever you want, but you will finally be thinking for yourself. You are ready!

While you may no longer trust the world around you (did you ever?), you now have an internal authority that is far more powerful and trustworthy. Turn inward to your own self rather than the feelings and perceptions of anyone else. Remember—they're all under influences of their own.

To become free from falling under the influence, you have to practice. Like anything else, it takes time and experience to get good at it. Get ready to disagree with someone. Be excited to say no. Look forward to leaving the group. Eventually it will stop feeling uncomfortable and start to feel normal. And when you slip up (we all do), don't be so hard on yourself; instead, embrace the fact that you are never finished, nothing is perfect, and nothing lasts forever. Remember that those wielding influence over you have a lot to gain and a lot to lose. Their resources and powers are becoming more invasive and aggressive every day. That's okay.

The greatest power is now in your hands—just knowing they are doing it.

Now it's your turn to raise hell, kids.

Go forth and fuck it up beautifully.

THE "PLANT YOUR FLAG" PRACTICE

The word *influence* literally means to flow in Latin.

It's a perfect visual for thinking about rising above the influence. The truth is, it's hard to know where to begin when influence feels like so much amorphous nothingness. When everything is everywhere all the time, we almost forget it exists, that it is already a part of us. Like stardust, silent bits and pieces come together over the course of our lives to build who we are at this moment. Considering it can feel overwhelming, like gazing into deep space. But if we remove the mystique of influence and begin to visualize the words, opinions, and ideas of others "flowing in," it makes it easier to conceptualize. It is also the very important first step in becoming uninfluenced.

The French took it one step further. Their definition of *influence*, from thirteenth-century old French, is "emanation from the stars affecting one's fate." While they meant actual stars, and not celebrities, the image of influence as an ethereal power that can change character or destiny is quite spot-on in both examples. Astrology or fame, it's of no matter. All that matters is the realization that someone or something is always influencing our thoughts, our way of living, our purchases, our behaviors, and our values.

Even once you become aware of the omnipresence of influence (*Outer World*), it's still happening around you (*Surface World*), and its impact still lingers from within you (*Inner World*) from before your great awakening. Maybe you're thinking, "*What's the point then?*" as if this book has been a lesson in futility. If you stop here, then maybe it is. But realizing who you are today as a direct result of all the flowing in since you were born is where you begin to take back your power.

Once you know that, see that, believe that, you have established your baseline. You can now choose what to do with the influence you've already experienced, how you'll receive it moving forward, and how to uncover who you are with all the layers of influence peeled away.

I am not going to lie to you—it is difficult to uncover who we really are when we've been receiving the flow for so long. Living unknowingly under the influence has profound, sometimes lasting effects. But it's okay. Below it all, you are there. And I am there, I promise. We still exist under the layers and layers of influence. The true you is that little voice that got quieter but was never fully silenced. It's that gnawing feeling. That tingling sensation. Your inner self is below all the muck, waiting for you to rediscover you. You can begin to crack open the nesting dolls of influence to locate your rooted, unwavering sense of self. In there, at the last casing before your core, you will find your free will. It's based upon the core itself—what you value. It's what makes up your character and lights up your soul.

And, of course, what we value is also based upon the many things that have influenced us, especially early on in our lives. As we learned in Part I, where we were born, the beliefs that were instilled in us, what we were lacking, what motivates us, and on and on and on makes us who we are—but that doesn't mean that understanding your core values (even if based on the influence of others) isn't important. It is. Incredibly so.

But I've discovered that it should really be a three-step approach to shake off values we've osmosed versus chosen:

1. Determine what you value.
2. Determine why you value it.
3. Adjust your value if the why isn't a good enough reason.

Knowing *why* you value something allows you to determine whether this is an original, deeply held belief that was chosen of your own volition, or whether it's been flowed into you. Uncovering our values without looking at the *why* behind them leaves us with flawed results, incomplete data. It's seeing the stars without understanding that there's a larger universe. Once we understand the root of our belief systems, then we can fine-tune as we see fit, getting to the truer nature of who we are, or whom we want to become. And that is what we stake claim to.

Historically, in our culture, planting a flag has symbolized ownership, possession, and colonialism. It's been used to make statements: "We were here" on the moon and "We aren't leaving" at the National Mall. This method of claiming, positioning, and firmly standing might not always have been used in the best way in the past, but I can think of no better way to use it than coming home to ourselves. We plant our own flags as we retake possession of ourselves, ownership of our lives, and stand firmly on our own two feet. Discovering who we are at our core (what we value) allows us to know what we must protect at all costs. You may think you value nice clothing, for example, or the way you present yourself. But then, after further investigation, you may realize that your mother is actually the one who cared about all that, or Instagram told you that you'd better be on trend—and now you recognize that you're actually good with a more classic wardrobe. Like the conquering of the moon, without remaining tethered to who we are, reaching for the proverbial stars can leave us floating in space.

Similar to the visual of flowing in, I love the one of a planted flag as a mental image of my own self-determination. This image first came to me when I was journaling on a much-needed trip away from the hustle and bustle of my life, and I realized that one of the values people often put on me, "achievement," wasn't a value of mine at all. I was taking on influence! The stake in the ground, my flag, is my freedom and independence. It's my strong will and autonomy. In a literal sense, putting a stake in the ground is used in reference to mooring or tethering something. By picturing my truly held values, convictions, and beliefs as affixed, it makes it easier for me to maintain an unwillingness to bend to whichever way the winds of influence try to blow.

Once we have found ourselves, and staked claim to that sense of being, it takes tremendous amounts of courage to hold true and firm when the world is trying to sway us otherwise. When we are tethered to our own sense of being is when we feel most grounded. Becoming unwavering and unyielding about our own convictions should feel solid, honest, and clear. If you were to think of times when you've taken the wrong path or gotten offtrack (like me with ambition!), I would guess that you were untethered from your glorious self either because you didn't know who that was yet, or you did, and you didn't protect it. Being brave enough to defend your core, your stake in the ground, not only keeps you *you,* but it also opens the universe to your most wild, and natural, reach-for-the-stars possibilities. It is here where we become free—and also very afraid. Surrendering to our deepest desires opens us up, but with that freedom comes vulnerability (the real kind). When the stakes are so high, it upends our previously held convictions, plans, and beliefs.

All those years ago, when I realized I was totally offtrack, it was the act of a simple handwritten list that allowed me to start to answer the question who would I be if no one was telling me who to be? Who was the real me if no one was watching? First, I drew a line down the middle of a sheet of paper. Above the top left column, I wrote the word *Less* and in the right column, I wrote *More.* What came out was surprising both in its simplicity and its radical honesty.

I wanted fewer work emails, sad desk salads, fewer gossip-based conversations, fewer alarm clocks, and fewer airplane trips, to-do lists, hellish relationships, and toxic people. I also wanted less monotony and routine, less stress and do-ism, less typing and watching, less noise pollution, less relentless communication, less achieving, less alcohol-based socializing, less swallowing my voice, and less people pleasing. And just as important, I wanted my shoulder to be less available to lean on.

The More column went on for pages. I wanted more freedom, more time, more family, more levity, more nature, more stars, more dancing, more love, more inspiring conversations, more make-outs, more creativity, more looking at the moon, more kids, more experiences, more books, more dogs, more quiet times, more horizons, more unknowns, more writing, more handholding, more laughing, more health, more life.

It wasn't that I didn't necessarily have those things—I often did!—

but it was the stuff that filled me up the most. With my Less list tipping the scales, it was leading me to want more of the More! Rightfully so.

If you only do this one singular and simple prompt, that is enough. It will show you how far you are from yourself.

Get out a piece of paper and write. Don't think. Just let it dump.

Recently I taught my son Hays about stream of consciousness writing. When you write that way, you do it to ignore punctuation, style, grammar, format—anything that stops you from writing. In fact, you shouldn't even lift your pen or pencil from the page. Even that break is a step too far. The idea behind this practice is to achieve a state of flow. It also helps with releasing bothersome thoughts from your mind and onto paper to allow space for real thoughts (like creativity or values), to have room to grow. When I taught my son about it, he latched onto the idea of dumping less useful thoughts so they didn't take up space. He will tell me he needs a pen and paper to do a "trash dump" of his mind before bed, before homework, or before leaving my house to go to his dad's. What a gift it has been!

So when you write your list, I want you to write it à la Hays. Remove the perfection of formed sentences and punctuation. Remove judgment and assessment. Don't think. Let your subconscious bubble through the rubble and guide your pen. You just might surprise yourself.

The key is to keep it simple. Don't worry about tools to write it or any rituals around it. Sometimes waiting for the perfect moment or piece of paper is just the excuse we need to never get started. Flow can happen anywhere—the back of a receipt while waiting in a drive-thru, a napkin in a bar, the inside of a book. You can zoom out or in on your list to focus on your life as a whole (as I did) or something specific like a relationship, job, or a big decision. Don't overthink it. Take as long or as little as you want. Just see what comes out when you are free from all the punctuation and perfect journaling. Once you have your Less/More list, hang it somewhere you can see it every day. Or if you don't want others to see it for whatever reason, hide it under your pillow or on the inside of a cabinet you open often. One time I found one I'd made for relationships. It was like half exorcism/half conjuring. Holding it up all those years later, forgetting I had even written it, to see that my relationship matched every one of those conjured wishes was super grounding. And very dope! I did it! The Less/More list can change with the tides of your life—new jobs, building a family, list-lessness are all reasons to check in—so I encourage you to revisit it often. Your values though, while they shift, often usually come back to the same overarching themes. Freedom as a core value might symbol-ize travel or financial independence, but becoming a parent might bring forth thoughts about having time for yourself. In both instances, the sense of freedom or autonomy is there, though in different forms.

In defining your values by observing them through the Less/More list, you create space to take a pause and make sure how you are spend-ing your time, your money, your life is in line with what you truly want and what truly brings you pleasure, joy, and contentment. It also shows you what you *don't* value, which is equally important. Values in practice create clear direction, choice, and action for where we want to go, who we want to be, and the legacy we craft. Our core values be-come our no-bullshit, unabashed, sturdy base upon which we will exist influence-free in the world.

Below is a list of core values commonly used by leadership insti-tutes and programs. This list is not exhaustive, but it will give you an idea of some common core values that people identify. And in the spirit of raising hell, by all means, add your own to the list.

Core Values List:

Achievement, Adventure, Authenticity, Authority, Autonomy, Balance, Beauty, Boldness, Challenge, Citizenship, Commu-nity, Compassion, Competency, Contribution, Creativity, Curiosity, Determination, Fairness, Faith, Fame, Friendships, Fun, Growth, Happiness, Honesty, Humor, Influence, Inner Harmony, Justice, Kindness, Knowledge, Leadership, Learn-ing, Love, Loyalty, Meaningful Work, Openness, Optimism, Peace, Pleasure, Poise, Popularity, Recognition, Religion, Rep-utation, Respect, Responsibility, Security, Self-Respect, Ser-vice, Spirituality, Stability, Status, Success, Trustworthiness, Wealth, Wisdom

Start by circling twenty that resonate with you. Go ahead, grab a pen right now and start circling right here right now, right in this book! Be careful not to choose things you aspire to be. While I would love to circle "Stability" because in theory I would love to have it, anytime I have had stability, I ran from it. Be honest with yourself. Once you have your twenty, cut it down to ten. Yup! It's hard! Now, do it again. Having five core values to focus on shows you what is really important to you, no matter how many screens or friends or cultural trends are telling you otherwise. If everything is valuable, then nothing is ever really prioritized.

On either the twenty, ten, or five list, now add the question Why? after each one. Keep answering and adding Why? until there are no more answers. What is it *about* that thing that you really want? For example, if you want more success, then what does it mean? Does it mean you want more accolades or outside validations (because those obviously don't actually do anything to create happiness)? *Or* is it about what success gives you—less financial stress, more space to work on your own personal projects, etc. And then you get to the bottom of what *kind* of success you actually want and what it means to you. This gets to the truth, the core, and allows you to move some values off your list if they are, in fact, ones that flowed in without your knowledge.

In practice it might look like this when picking apart a value you circled:

Popularity → Why? → It means I'm liked → Why? → Because I am nice →Why? → So I can have friends → Why? → So I have people to do things with → Why? → To share things I like → Why? → So I won't be alone → Why? → Life is meant to be shared → Why? → Because loneliness means I am not loved → Why? → Without being loved, I don't belong.

Instead of "Popularity," maybe "Belonging" is what should be on your list.

Belonging as a core value is very different from popularity. Once we look under the hood and get to the core of our values, we can build

our lives accordingly, more authentically. Someone who circled popularity without ever thinking about why they might value it could get
caught up with people who simply stroke their ego, when in reality
what they really were seeking, and what their value likely is, is to be
seen, understood, appreciated, and included.

Now, what are your five? Whittle down the why's until you get to
your core values. You can work on them here:

In becoming more aware of what has flowed in from elsewhere,
we'll be able to shed those ideas and inclinations that are not aligned
with our values. Personal values shift and change based upon experiences. Having kids, meeting someone new, moving abroad, learning
something new (like this exercise!) can radically transform perspectives and behaviors, taking lives in unknown directions. All this is to
say that your list, though moored/tethered/staked, is yours and yours
alone, but it should be revisited from time to time. It's not a lasting
conclusion, as your life, based on new thoughts and new eyes from
new experiences, will constantly be evolving. And because what is
most valuable to us is also what's most susceptible to falling prey to
influence, we must always relook at ourselves and guard what we discover with care.

The first step in becoming free from the influences of our world is
changing how we understand ourselves. Instead of breathing in the
Folktales and the *Folklores* and the phrases that create our identities,

we can establish our own, to fill our heads and worlds with our own voices.

We are not finite. Much like the universe, we are expansive. How wonderful is that?

Folklore   You should know what you want.

Folktale   If I don't know what I want, I am without purpose.

Folk Rebellion   Knowing what I want takes time, intention, and work. It also changes, but I will revisit my values to make sure I am not offtrack, and if I am, get back on.

THE "PEEL THE ONION" TECHNIQUE

Look at you, big shot. Now you know who you are. Or you're getting there. You might be sitting a little taller now that you've got a handle on who you are, your values, and what you must protect. It's a great place to be. Discovering a true sense of self, free from influence, is wildly empowering! But, like a fake life online, it can be hard to maintain.

This begs the question: How do we protect our true influence-free selves?

Nothing is wilier than influence that's been artfully crafted to flow into you without your even realizing it. As we now know, regardless of the topic, we are constantly being attacked by pervasive, insidious, and tricky outside influences with hidden agendas. Before you know it, your foundation of core values has been shifted like an unexpected rearranging of the furniture—all the pieces may be there but not quite how you left them. While the concept of influence has been around for centuries, psychologists put it under a microscope, ushering it into the public discourse in the 1950s with the phrase *subliminal persuasion*. In the seven decades since the term was coined, so much has changed. Those harnessing the influence have grown exponentially. What were once hidden intentions now pull up a chair and sit brashly beside us. Being skeptical about influence was a part of understanding it back then. But today influence is praised and accepted as a normal part of life. Professional "influencers" are put on pedestals and looked up to for guidance and trendsetting. Skepticism has taken a back seat to the acceptance of these practices, and as a result, we have entered an age where we know no different.

Ironically, our society has also grown more and more obsessed with exploring "the self," commodifying authenticity and identity to the point where they've become buzzwords without real meaning. Those very words and ideas have been co-opted by the now-common professional influencers who make it their job to proselytize about navel-gazing and living a "real" life using artificial images and videos of their "perfect" make-believe lives.

It all underlines how essential it is, now more than ever, to raise red flags, to question the status quo, and to trust our guts despite what our eyes might see. It is through skepticism that we find the key to keeping our true sense of self once we have uncovered it. By becoming skeptical, we are granted the ability to know from whom we must protect ourselves.

Discovering the source of who or what is trying to persuade us gives us the chance to attempt to understand why. The only way to unwrap the intent is to stop assuming everything we hear is true, everything done is with the best intentions, and everything believed is correct. Beliefs are not the same as facts. And uncovering the true motivation of sources is the best protection of all. We already know the many sources of influence that originate from our *Inner World, Surface World,* and *Outer World* spheres, but by drilling down to identify exactly which entity is behind a given influence, we can better uncover their motivations, and then, armed with that insight, make intentional choices true to ourselves and above the sway of their influence. I've been called a skeptic, a pessimist, and a doubter, but rather than be offended when people refer to me that way, I share that being those things makes me a researcher, questioner, and objector if needed. It's those labels that make me do what's imperative in rising above the influence: think critically.

In our sick society, it is healthy to use thoughtful reasoning when assessing just about everything—media, government, technology, societal norms, religion, politics, and big businesses like food, pharma, and finance, etc. In fact, that has proved to be my greatest asset for living a good and happy life. Living a life as a doubter is actually how we are truest to ourselves, and as a result, live the most fulfilled lives we desire.

Skepticism has led the way for my independent thinking. What started as a distrustful instinct and inclination toward curiosity has been cultivated into a thoughtful way of existing. In today's world, this has become a rare commodity, and I believe it will be key to bringing us out from under. The more we (targets) become knowledgeable about the manipulators, the better we can spot their efforts and refuse our cooperation. We won't be patsies anymore. We can, and will, rise above! But we have to choose to do the work.

My hope is that it becomes more commonplace (dare I say *mainstream*) to see the world through a questioning lens. In the past, when people remarked that I saw things differently, it would lead me to think about how and why. I kept coming back to my squinted side-eye—the skeptical lens through which I view most everything. Whether it was technology, the path to college, business, or relationships, I was able to see the bullshit through the trees. Coming out from under the spell of influence means those rose-colored glasses come off, the veil lifts. Once you can adopt that attitude, it's a mindset that can be applied to everything.

By becoming skeptical, stepping into the seat of thinking critically, we eject ourselves from the default mode. It is then that we begin to see the motivations of others.

I call the process of this uncovering the onion technique because there's an almost never-ending cycle of layers of influence to get to the core, which often stinks to high hell.

And it starts with asking questions (which you might not realize but you've been practicing every step of the way on your way to raising hell, through the *Raise Hell prompts. See what I did there?).

When we use the peel the onion technique, we drill down by taking the next step, and the next, and the next until we reach that crucial core:

Top of the Onion

Be Skeptical—*What is the motivation? Who wins, gains, benefits? What are the benefits to the winner? Who loses? What does the loser lose?*

Second Layer

Double Doubt—*Who else does this benefit? Who else does this affect? Who else is involved? What other systems are at play here?*

Third Layer

Find Facts—*What is fact versus opinion? What does the data say? How is the messaging slanted? Where does the data come from?*

Fourth Layer

Instinct Gut Check—*How do I feel? Does it appeal to our insecurities and anxieties? Is there a clear purpose to get us to respond emotionally?*

Fifth Layer

Curiosity—*What are the unknowns? What has history shown us? Does it use any means to an end? Does it aim to influence large populations toward some type of cause or position?*

The Core

The Truth—*Now that I have questioned and identified all of the factors of influence in play, what is the conclusion I can come to about what this influence truly wants from me? What are the big and small impacts of the truth if I choose to accept this influence—are they in line with my values? Does choosing this influence help me to protect my planted flag or threaten it? Would I prefer to say no to this influence and choose a different direction instead? What informed, thoughtful, intentional decision do I want to make now that I have uncovered all these layers?*

By asking questions like these, I can come to a conclusion about what's best for myself, my family, my health, my well-being, and my happiness, and protect my planted flag. Sure, I find pleasure in putting things under a microscope. But doing due diligence on apps, people, press, food, trends, sayings, beliefs, and religion shouldn't feel like a slog. Like a modern-day Sherlock Holmes, I let my curiosity free to roam all over, devouring all I can to make informed, intentional choices. Cutting through the noise to understand the answers to the questions above lets a person make solid decisions based on a well-rounded understanding. In the end, I simply think, *Prove it to me.* That's it. That's all you have to do.

People need help understanding that the motivation of others to influence is so strong that they put enormous amounts of effort and money into it. And with the ever-growing volume of content and platforms with less than noble intentions, how do we begin to help peel back the truth behind their motivations? A layered onion. That's how.

If we were to examine the news we receive, for example, our imaginary onion might look like this in practice:

Top of the onion: We may encounter the news via our social media feeds. When we are met with this influence, we must first **be skeptical**. Here we need to understand that this is a platform often used by non-mainstream media sources. The goal here is to create and drive alternative narratives. And if it is from the mainstream media, it most likely has been crafted to spark an intense feeling to get a person to click. The content is polarizing on purpose. Because of this and the unscrupulous entities often spreading misinformation here, it is wise for us to look for information elsewhere.

Second Layer: We then must peel deeper and **double doubt** the influence presented to us. Who else is involved in this influence? The people writing the articles have been hired to write for that specific publication. That publication has a set of beliefs, a culture, even core values that ladder up to the overall initiative and narrative of the organization that owns it. It is through that lens that the writers are often expected to write to continue to uplift the overarching narrative or remain on brand. The writers, employees, journalists, and content creators have a responsibility to their employers (clicks, subscriptions, visibility) and themselves (paychecks/success/promotions). Therefore, the articles are written under the influence of both the organization and in an ideal world, or at least in the best scenarios, the writers are given autonomy and are reporting in an unbiased way. If we look further still, we can realize that the overarching narrative of the entity is one that is driven by the leaders within the organization. The leaders of the business are chosen and placed based on systemic norms—leaving those positions to be filled with certain types of people, with certain types of goals and motivations themselves. The goals and motivations of the key leaders and executives within the organization create a hierarchy of conformity. Because of personal quests—success, money, notoriety, respect, etc.—those with influence in the organization will fall in line with what the organization wants while the non-compliants who speak up or step out of line are weeded out. Last, the organization is looking to do what is in the best interest of the company. In some cases, that means keeping stakeholders happy; in others, it's keeping the employees working and motivated; and in most, it means increasing profits above all else.

Third Layer: Now we can start to focus on **finding the facts**. When assessing the content of the story, we can look with a shrewd eye for supporting information, or information to debunk, on our own. Because even data can be coerced to fit a narrative or pulled from less than reputable sources (publicists, marketers, politicians, etc.), we know that analyzing the information shared both for its context and origin is a necessary next step. Learning the difference between opinion and reporting is up to us.

Fourth Layer: Now beyond wading through the opinions-versus-facts, we can check in with our **instincts and gut check** how we feel. Seeing the headline designed to spark our emotions through manipulative communications allows us to acknowledge how it could have made us feel if we'd let it. Was the intent to increase our feelings? If you feel anything other than informed, you've probably just been manipulated in some way—and now you know it.

Fifth Layer: This micromoment (your clicking this headline) might seem insignificant in the grand scheme, but **being curious** and asking more questions still might lead you to recognize the greater systems at play, like the shoppable product link to mask our feelings of inadequacy embedded in the last paragraph of the "article."

The Core: At the end you finally **get to the truth**. You'll acknowledge the institutions in the *Outer World* (economics, media, etc.), the isms and expectations in your *Surface World* (community expectations and standards), and see there is a call to action at the end of the article to put a hard press on your *Inner World* (how you feel, what you believe, etc.).

You decide the article is shit, and you don't need another facial serum to "be your best self." Maybe you take it a step further, raise some hell, and promise yourself never to read something from the entity trying to trick you again, and from here on out you'll take a hard look at how the structure of capitalism trickles down into everything, including how it makes you feel.

This is not to say that everything in life climbs up to the influence of capitalism, but I'm hard-pressed to do this exercise and not find money or ego waiting at the other end. The practice in this example helps us to see how it's possible that we are being manipulated into outrage, or why we can't stop clicking, or that a news organization is still just a business looking to remain relevant and profitable, often helmed by people driven by extrinsic motivations like personal success and financial gain.

The onion technique helps us to reach the core truth, and it can be applied in more ways than one. After we peel back the layers of the onions trying to influence us, we can begin to peel back the layers of our own actions as well. Under the surface, there's much to uncover. The overarching goal is never to simply accept what you're being told without considering the source, whether it's about the next great band (according to whom?) or our country's next leader. I could be talking about technology, the coronavirus, parenthood, finances, or the next newfangled wellness offering—it might stink. Once you figure out how to ask questions and sniff out the convoluted truth, even if it brings tears to your eyes, you can sniff out most everything for what it is.

Folklore   The surface is what is true.

Folktale   Whoever I see is the one who is pulling the strings.

Folk Rebellion   There are multiple layers to everything, and only by digging deeper will I be able to discern true intentions so that I can make my own intentional, uninfluenced choices.

THE "MIDDLE FISH" ASSESSMENT

Growing up, I desperately wanted to be a big fish. Once grown, I appreciated the freedom in being a small one.

But it wasn't until my wake-up later in life that I truly understood that we are neither big nor small. We're in between. Maybe this was implied, but I always pictured it as a binary option. It was lost on me that, by default, there were always more fish because there will always be a bigger fish, and always smaller ones. I now recognize that I am a bigger fish to others—even when I feel small. And when looking at the smaller fish, I see I am the smaller one to others—even when I feel big. We are forever the middle fish.

Imagine a picture of three fish in a row, mouths open, going from bigger to smaller, left to right. When I think about the influence that flows into me and the influence that flows out from me—I picture these three fish.

When assessing influence, we can turn to these three fish to see how being at the mercy of the big fish (influence flowing into us) will affect our lives, and those around us—the little fish (influence flowing out of us). We have long known that in the sea, and in life, the big fish consume the small. What we allow to consume us has the potential to change us, and then we pass that consuming on down.

When I was being eaten alive by the influence of productivity and grind culture, it flowed through me (the middle fish) and I was influencing everyone around me, from people who worked for me to my own family. I was letting them think they needed to work harder and longer, and I was normalizing being exhausted, burned-out, and overwhelmed to my kid.

Now that you know what you value (and why), and how to get to the root of the source of influence, the next step is to decide what to do with that knowledge. Earlier, I shared with you the concept that when given a choice, choose the path that makes your life expand, not diminish. But that was before we knew how to weigh those choices against our values and how to look at them through the layers of influence. The middle fish assessment is where you calculate if what is trying to consume you not only makes your life expand—but taking it one step further to see how it will flow out of you, what the trade-off is on the other side.

When I consider bringing a new piece of technology into my home (yes, I do this often, consider things before acting, and now you will too!), I weigh it against my beliefs and values, dig into all the influences at play, and if those check out, I then think about what this influence is going to do for me; I consider how it will shape my life but also if it will expand or diminish down the line, outside of myself.

My middle fish assessment process looks like this:

SMART DEVICE	ME	KIDS
Hands-free	Automated	Doesn't think for self
Advertisements	Buy things	Future consumer
Convenience	Free time	Develop friendship
Listening	Loss of privacy	No boundaries
Learning	Can't control	Data collection
Many uses	Cannibalize IRL	Steals hobbies

When I do this assessment, I can see the good, the bad, the trade-offs and then make a decision. While I very much would like a reminder to bring an umbrella from time to time so I don't get stuck in the rain, the trade-off for that convenience is not enough for me. I can see down the line how my lack of control of the algorithms, data collection, and advertisements will foster an overreliance on devices such as this for my kids in the future, grooming them to become good consumers who don't think for themselves, have no boundaries, and think that a listening robot who helps them, but makes that trade to learn from them and about them, is normal. It's not. So as much as everyone tries to influence me otherwise—an Alexa will not be becoming a member of our household.

With this assessment you can see how much larger than life that big fish is and cut off their feeding frenzy.

Folklore   There are only big and little fish.

Folktale   Influence stops at me.

Folk Rebellion   I can see the influence of my experiences and choices flowing through me into the world and make intentional choices about which big fish I allow to influence me, moving forward with understanding my place in the tides of life.

THE "SNOWMAN" EFFECT

The *folklore* of my birth, as told by my family, is that my arrival was almost by way of a snowmobile. Either a doctor arriving on one to deliver me at home or, a funnier image, a neighbor picking up my very pregnant, two-weeks-overdue mother to sidesaddle her belly to the hospital.

As legend goes, it was Christmas Eve, 1978, and my family was ensconced at my grandmother's house, playing cards over cocktails. With the curtains drawn on the big picture window, nobody was aware of what was happening outside—too busy focusing on their bets and mostly disinterested in the weather because a little flurry in Upstate New York was a normal part of life.

Unbeknownst to everyone, including the weathermen, a major unexpected winter storm was blowing in, sneaking up from the south. After the fact, as with all upstate winter storms, there would be squabbles over who accrued the most inches. We later learned that I was born in the epicenter, with our snowbelt communities reporting upward of thirty-five inches.

Snow like this wasn't something we were afraid of. It wasn't a reason to stay indoors. It was rarely something that evoked panic buying. We even drove in it. Learning to drive while it snows was basically imperative if you ever wanted to leave the house from November to

April. Some mornings you would wake up to doors that wouldn't open. Roads could look like canyons in February because the snowbanks were so high. Digging out was commonplace. When we talked about the weather after the fact (and boy, did everyone love to!), it was with astonishment and wonder, chuckles and head shakes. I don't recall, "It's going to snow in three days! Might be a big one!" We never looked forward like that. If a storm was coming, it was not really a big deal and not predicted that far in the future.

I know I say this a lot—but things were much different back then. As I sit here writing this, the New York City school system has declared snow days a thing of the past. With Chancellor David Banks gleefully stating, "*There are technically no more snow days . . . with the new technology that we have—that's one of the good things that came out of COVID—if a snow day comes around, we want to make sure that our kids continue to learn. So, sorry kids! No more snow days, but it's gonna be good for you!*"

There are so many things wrong with this hugely influential statement that it's exhausting just to think about picking it apart. Like a narcissist's word salad or the state of the internet—when everything is wrong, false, skewed, or covertly implied—we can be overwhelmed just by the choice of which part to correct first, often pushing us to choose the simplest option, thinking *it is what it is,* shrugging our shoulders, and doing nothing.

But we aren't going to do that now, are we?

To rise above the influence, we must do the work to pull apart the many influential factors at play here. This is where we bring all our tactics together—but in the reverse order from how we learned them. The influences of our *Inner, Surface,* and *Outer Worlds* all ball up into one. Like building a snowman, the largest structures in our *Outer World* build the foundation, which is why we look to middle fish first. Next up, in the middle is our own sniff test on how this will show up in our *Surface World,* and so we must peel the onion. And like all good snowmen and snowwomen, our crowning glory of beliefs, values, and thoughts resides in the head, our *Inner World.* And so we plant the flag to top off our snowballed-up being of influence.

Let's begin.

Step 1: Middle Fish

SNOW   ME   KIDS

Media	Fear of snow	Fear of any weather
Apps	Live tracking	Overdependence/unreliable
Snow	Makes me happy	Missed experiences
Apparel	Winter coat, boots	No bad weather/bad clothes
School	Zoom snow days	Classroom only place to learn

Step 2: Peel the Onion of "No more snow days."

Top of the Onion

Be Skeptical—*Do the people who this will affect—students, teachers, and parents—also think that this is good? Good for what? Good for whom? What is the plan for protecting our children from the harmful effects of being on more technology?*

Second Layer

Double Doubt—*Who benefits? If I follow the money, where will it lead? What's lost by this? Do the students understand how the technology works, who it benefits, and how it's made?*

Third Layer

Find Facts—*To which technology is he referring? What is that technology replacing? How does it affect the experience of the teaching? What research has been done on this technology? Have students of all ages taken a digital literacy and citizenship class? Have the apps, programs, and ed tech all been vetted by experts on privacy, safety, and online communities? Does every child even have access to this technology, regardless of their family's socioeconomic status? Are the technologies gaining the data of our children? How does this technology affect the experience of learning? What is this technology shown to do well? What is the process for setting up safe accounts and log-ons and keeping them safe?*

Fourth Layer

Instinct Gut Check—*Does this make me feel like the kids will expand or diminish? Do I feel concerned? Uncomfortable?*

Fifth Layer

Curiosity—*Does this perpetuate inequity? Has the number of hours a student will be expected to be on the technologies been compared* *with the research at which it starts to negatively affect their posture, mental health, degenerative myopia, nearsightedness, connection to others, and cognition? What did we lose by having the teachers spend more time being IT professionals than teaching?*

The Core

The Truth—*Was this a decision influenced by outside influences such as politicians, money, technology, and ease without properly understanding the consequences of such an influential decision?*

Step 3: Plant Your Flag

Pull out your list of values and the flag planted beside it. Ask yourself, "Which of these values does this proposal agree with, if any? Which is it in conflict with, if any?" Determine if the proposal is in line with your values and whether you need to plant your flag to support or defend them.

This is what it looked like in practice for me. Here is how I assessed the influence of this proposal using the tools above.

As a mom very connected to snow days, I'm thinking about where this influential decision in my child's life will flow and what it will leave in its path. I can see in my middle fish that it will affect everything from how my kids perceive the weather to how they learn, and how I handle fear and risks and technology. This ball of influence can be immense and far-reaching if I allow it.

To understand how we arrived here, with banned snow days, we peel back the layers of the stinky onion and see more deeply what is the true core motivation of the New York City Department of Education.

First, I'm skeptical about the mayor's assertion that this proposal is "good." There's the question of what it means for something to be "good for you." After all, while there is endless data now about the negative effects of tech overuse on adults and children, there are countless studies and research projects dedicated to the positive effects of both play and exercise for kids and of spending time outdoors and especially in nature, leaving me suspicious of this claim of "good." By these facts alone, it could be said that a day spent sledding is, arguably, much healthier and more beneficial for a child than a day spent staring at a screen and *perhaps* learning something. So if this decision wasn't good for the kids, who was it good for?

The influence of unpredictable weather, and the opportunity for every person to overanalyze and obsess over our unpredictable weather via technology, isn't great for school administrators or mayors. When it comes to running the largest school system in America, you're damned if you do and damned if you don't. Closing schools over the alarm of an impending storm is criticized after the fact, when not enough snow fell to shovel. When the opposite happens, and a key leader decides a little snow never hurt anyone, leaving schools open while a lot of snow accumulates and makes it hard and risky to commute home, they're scorned for allowing our precious snowflakes to become subject to some. Polarizing hot takes around equitability—socioeconomics, ableism, privilege, etc.—creates infighting rather than equitable solutions. Snow days may not be great for the approval ratings of those in charge, but the gnarly weather has become really beneficial to another very influential bunch: the news media.

Ratcheting up sensationalism has the headlines of impending inclement weather starting earlier than ever before. Blips on the radar that won't *potentially* arrive on your doorstep for half a week become the thing to feed the machines of clicked links and audiences of potential "storm forecasted" very far in advance. The earlier the news platforms start, the more they cover it, and the more anticipatory anxiety is created.

Depending on the size of the storm or the size of the words used to describe it (or the volume of the ominous music playing in the background of the storm watch promo—like some kind of disaster movie trailer), you might call the anxiety *panic*. New words, with the goal to invoke fear, are created, one bigger and badder than the next because we are all becoming anesthetized to the sensationalism of the marketing of weather. A blizzard became snowpocalypse. Snowpocalypse turned into snowmageddon. And then—*dun, dun, duuun*—bomb cyclones. The naming of winter storms, as the National Weather Service does for hurricanes, is something fairly new. The Weather Channel decided to roll that out for the winter of 2012–13, claiming it was to help with storm awareness and preparation. In reality, it's a gimmick by a media organization to drive big business through people tuning in to their station because the actual National Weather Service does not name winter storms. Assigning names and airing dire warnings on TV to "prepare" for every system that passes through takes away from the flagging of actual natural disasters when they do happen. It also has, like most media, desensitized us by being too much, too often, too extra.

Under capitalism, the media companies want more. More opportunities for audience attention to sell ads. The result is often *less,* and they end up cannibalizing themselves. Like a restaurant that has franchised too fast or eating the same food for dinner every night, what was once special has become commonplace. Background noise. Static. Humans become disinterested and dispassionate when the tap is left on. And yet the prepared-panic influence still trickles down to the good consumers doing their duty of flocking to grocery stores and gas stations. As the news spreads, spending increases.

So does the need for childcare to alleviate the stress and anxiety of parents now tapped with having not just children home but children home who have to be taught, on a screen, while the flakes fly outside. Children who are a part of the generation with rising anxiety who could very much benefit from unstructured outdoor play.

Weather forecasting and weather itself have changed significantly as technology has improved and our climate has worsened. While predicting storms in the past was less exact, the expectations were, as such, based on that knowledge. It wasn't an exact science. Predicting it today, with all the innovation, satellite images, radar, and live apps telling us if we should bring an umbrella, has conditioned us to believe that it *is* more exact. Unfortunately, and ironically, because of our own progress, the weather has become more unpredictable, regardless of the tools in place with which we can watch it. That doesn't change wanting to know, needing to know, or expecting to know if the weather is going to rain out your camping trip. How many of us have actually stood in the rain, looked at our weather apps, and said, "But it says it's not raining!" Like much in technology, becoming so dependent on it takes away our free will, like mindlessly (or dutifully) following Waze off a cliff. Some of the best vacations I've had were ones where a weather app had predicted a 50 percent chance of rain in a popular travel spot. Opening the weather app and looking at the ten-day fore-cast and seeing lightning bolts over a three-day weekend has hordes of people canceling their plans, not wanting to waste the money to sit inside. Now, I know, because I grew up without this app, that the weather often changes. I rarely look at the apps, but when I do, I take them with a grain of salt. This also might be why I am often left caught in the rain without an umbrella, but it's a small price to pay for not ob-sessing over something I cannot change—but could drastically change how I make my decisions.

Decisions. Free will.

This is what it all comes down to.

What will you do with all your newfound clarity and knowledge? Now that you've got your answers, your awareness of how this influ-ence could flow in and out of you and how it would live beside your values, you have a choice to make.

For me, in this instance, I can see it is pulling me, and my children, from my stake in the ground—my flag.

The very idea that there is no value in learning outside a classroom goes against everything I believe. My core values dance around inde-pendence, curiosity, and nature.

My skepticism has led the way to my understanding the incredible benefits of learning through play, being outdoors, and creating com-munity. Reading the endless data about the negative effects of tech and through my own curiosity and refusal to take things at face value, I know the answers to all those questions regarding the technology that I posed above.

So for all these reasons, after some serious consideration and criti-cal thinking, I will be opting my children out of remote learning on Zoom snow days in perpetuity because my Snowman that I built (mid-dle fish, peel the onion, plant your flag) said so. You can find my chil-dren, cookie sheet in hand, lumbering through the snow to sled down any hill/ramp/dog park they can find for all future snow days. My only concern is how long it will be before some well-meaning person de-cides to call child protective services for my kids' absence from school in favor of wandering the snowy streets in search of a childhood rite of passage I intentionally chose for them based on research, science, and values—not influence.

Setting aside my nostalgic whimsical ideals around snowball fights and hot cocoa–filled reading nooks, and the magic of childhood, I've already made my decision of what I will be doing under this undue influence, based on my information and sense of self. Now I can turn to see what I want to do about it.

Make trouble? Raise hell?

You bet your ass.

Folklore Life, like weather, is unpredictable.

Folktale I can be prepared for anything life (or weather) throws at me.

Folk Rebellion We can deal with influence in our lives, even when unpredictable. Using my *Folk Rebellion* tool kit, I know I can confront the unexpected influences that pop up in my life and ensure I am making the intentional choices that are true to me, raise a little hell by stepping out of line, and truly live well.

EPILOGUE

(or a Letter to You All About the Influences of this Book & Me)

This book is a culmination of all the things in my life that have ever influenced me coming together to make me me.

Only I could make this book.

It's based on an indescribable amount of things, moments, memories, motivations, beliefs, opportunities, experiences, people, hardships, wins, losses, loves, trends, information, books, learnings, education, economics, privilege, pivots, personalities, mindsets, facts, and so much more. But they are mine.

That is why every book is different. Every lens is unique.

I wanted to wait till the very end to tell you this so that it wouldn't influence your opinion of me, the writing, my parenting, or my expertise as your guide, but I wrote this book during the hardest year of my life. I've had sadder ones, more gutting ones, busier ones, but this year was by far the most difficult. Truthfully, it put me into a vise and squeezed the ever-loving shit out of me.

This fact would result in me shaking my fists toward the sky lamenting, "If only I were a man" while two children clung to me for their every need. I'd sob, "Why fucking now?!" with regard to my lifelong dream of becoming a published author being granted while crying from a hospital room as I watched over my baby recovering from his second open-heart surgery in four months. I'd scream, "I quit! I can't fucking do this anymore!" when a new travesty occurred in the world, redirecting me from my craft—school lock-downs, subway bombs, baby formula shortage, etc. All I could think was what would it be like to write a book without all the trauma and drama? My husband, Mikey, would let me have my pity party, some space, and then talk me off the ledge, bring me M&M's (dear God, I miss cigarettes), and help me reframe.

Those *Inner, Surface,* and *Outer World* influences made this book what it is.

If I were a man, I would not have written all the essays on gender, pay disparity, or noted unconscious bias. If I wasn't a mother under duress with her child in the riskiest of situations, I wouldn't have been able to write from both sides about risk versus reward. And if I wasn't living in the United States during what I very much believe is a dystopian nightmare come true, where laws keep being passed to protect businesses and profit, instead of my body and my children's ability to safely go to school, I wouldn't have so confidently and angrily waded into the polarizing gray area discussions on capitalism.

Once again, the juice was worth the squeeze.

And when I think about all the tough stuff that went into the building of this life's work, I'd be remiss not to note the good things that allowed me to still get it done despite all odds. Things that I realize influenced my ability to do it when others might not be so lucky.

I have a loving and supportive husband who could and did take over for the better part of a year. I have parents and in-laws who stayed with us to help raise the kids while I toiled away. We had the means to hire additional childcare to help when I felt like we were about to finally crumble. We live in a safe place. I have a roof over my head. My kids are goddamned warriors.

All this is to say that I've thought a lot about the exhaustive list of things that make me me and made this book what it is. I thought it important to share them, so nobody is under any false pretenses of my biases, my influences, my privileges, and my offerings to you. While this is not a complete account (we'd be here all year), it shares some of the more poignant themes.

Inner World

I am the oldest of four siblings, a child of divorce, and currently forty-four years old and (basically) healthy. I am Caucasian, a conventionally pretty female, half Irish and half Italian, born and raised in Upstate New York by (at the time) economically lower-middle-class parents, and came of age in the eighties and nineties. Currently I am economically upper-middle-class but regularly have insufficient fund notices for prioritizing my art over my income, the high cost of living in New York City, and the debts I am still paying off that I accrued over my divorce. I'm of Gen X and the sandwich generation. I was loved growing up. I am loved now. I'm a voracious reader. My interests swing wildly based on whatever door (book) I opened that week. I am fueled by coffee, culture, music, stories, and being a good mother. I believe stupidity is the greatest sin of all. Change is something I welcome. I am rarely afraid. I always figure it out. While spiritual, I never bought into the structure of organized religion, but I believe in the stars. People assume I am extroverted, but I am not. I am a highly sensitive person, an empath, an introvert with off-the-charts Inattentive ADHD. My Myers-Briggs is INFP. I was addicted to cigarettes many years ago, and I miss them all the time. I was pregnant when I sold this book.

Surface World

I live in one of the most desired neighborhoods in Brooklyn. It happened accidentally; I constantly think I don't belong here and do everything I can to keep the apartment, which is well above my financial abilities but good for my kids. This proximity to wealth and writers has helped me and hurt me. My days revolve around my children, filled with drop-offs and pickups and meeting their needs, and splitting time with my ex-husband, who lives (for better or worse) two blocks away. I got remarried this year despite swearing off marriage to a first-rate amazing man, a friend for fifteen years who changed my *Inner* and *Surface Worlds* so much for the better that it changed my views on the *Outer World* influence of matrimony! I work from home, which used to be good for creativity but is now bad for focus with a husband and kids around. And the dog. Kylian is my giant Newfoundland who sits under my feet and is afraid of the world. His barking takes me off task. My community has grown up as I have. I am surrounded by people who can help me, advise me, connect me, guide me. There is a multitude of things in my *Surface World* that make me a very lucky person. Also, the nearness of the best coffee shops and baristas and pizza. While writing this, my *Surface World* shifted from my neighborhood to a hospital room where I lived with Beau for three months at Hassenfeld Children's Hospital.

Outer World

I am a mother raising boys in a world that is still taking away their mothers' autonomy. I fear for their life being taken by a gun every time I drop them off at school or visit a public place. Because of my grit and hustle from the cultures I became a young adult in, I am guided in the belief that I can change things, fix things. It's the time of the digital revolution and technology is all around us, though I try my best not to let it pervade every inch of my life. I consume media in 3D format whenever I can and avoid most of the social media, reality TV, and popcorn content of today's culture. I've voted for both Republicans and Democrats at differing times. The politics of today has turned me sour on the system as a whole. I live in a capitalist and patriarchal system in New York City, in Kings County, in New York State, in the United States of America, and exist within the established structures and systems of those jurisdictions. I am a human of the world, which is dying because of climate change.

See how that works?
Now go do you.

Acknowledgments

To those who got me over the finish line.

Mikey,

My love. Without you, there would be no book to give thanks in. As I write this, you've loaded the dishwasher, made me coffee, and changed Beau's diaper. You tell me "you got this babe" every time I think I don't. You remind me of our far-off dream and what all this is working toward. For all the times you sat and let me read out loud "just one more page," for the beautiful places you took me to that I got to write from, for the sandwiches you brought me when I refused to come downstairs, for all the games of basketball you played with Hays. I am eternally grateful. Most important, thank you for trusting in me when I said I was no longer going to do marketing or branding. Because of your support, I was able to turn away from paid work and make this come true. You allowed me the opportunity to realize my dreams. I know I owe you some dinners out, but who's keeping score? Hopefully this is a bestseller and I can get you back with a cottage in a seaside village where we can grow old and everyone leaves us alone. Wait. Was this the long con? I am so incredibly grateful for your partnership in our colorful life. I love you so much. How lucky are we?

Hays,

Thank you for making me take breaks to play basketball with you or jump in the pool "just quickly" on vacation. Thank you for understanding when I had to take my laptop out but didn't want to. Thank you for recognizing I was tired or stressed and in need of an easy bedtime without the games. Thank you for telling your friends your mom's "an author and her book is gonna be in Books Are Magic and it has a swear word on the cover!" with such pride. Thank you for recognizing when I haven't had my coffee yet. Thanks for saying "yes!" happily, every time I said "can you please sit with Beau" while I tried to finish a sentence. Thank you for dancing with me when I sold the book, when I submitted my drafts, when it went up online. Thank you for saying "We don't keep it in, Mom" and then taking me through your calming exercise, tracing up and down your fingers while I took a breath. Thank you for the homemade countdown you made for our fridge calculating how many words I needed to write each day to meet my deadline (that wasn't anxiety-inducing at all)! Thank you for the ten-second heart hugs and the "You can do it, Mom!" when I swore that I couldn't. But mostly, thank you for being you—for our walks and talks and bedtime chats about life and the stars and what's important and what's fair and what's not and how hard and amazing it all is—which have been the source of inspiration for this book and many of the essays within it. Your brilliant mind, keen eye, and deeply feeling soul keep me on my toes. I promise to keep trying to make sense of it all for you. I love you more than the moon and the stars and the whole-wide universe, lovebug. Now, go out there and raise some hell, kiddo.

Naa-Naa & Karen,

Thank you for stepping in for me and being my Beau-bear's childcare while I slunk into the corners of the house to write. Thank you for giving grace to the crazy person in pajamas who only came out for coffee or a crying baby! You both have made this book possible by extending your time, love, and understanding to Beau and our family. I would have been lost without you!

Readers,

I haven't met you yet, but I've pictured you a thousand times. Like right now. It's three A.M. and past the deadline and my eyes are melting down my face and I want to close the computer and be done, but then I picture you reading this book and I know I can push it just a little further. Thank you for being my muse.

To those who brought this book into the world.
The people who believed in me, my voice, and my idea long before it
became a tangible thing of paper and ink.

Lucinda Halpern,
I will never forget our first conversation. I was looking for a literary
agent who would "get it" after many others wanted to change the title,
the concept, the structure to something more easily sellable in the
market. You saw the "Big Idea" and the opportunity if done right.
Thank you for holding my hand through the process of the proposal,
believing in me, and standing behind my vision. There are no words to
say how much you changed my life when you sold my book. You took
a chance on a cold email from a bullshit artist and turned her into a
published author and for that I am eternally grateful.

Nora Zelevansky,
The synthesizer of my madness. When I thought there was just a
bunch of random half-written stories, you found the scaffolding. I
don't know what my writing would be without your sage wisdom,
discerning eye, and coaxing out and making sense of "what I am trying to say." Thank you for being my sanity, my confidante, my writing
buddy. You saved me from the isolation of a writer's life by riding
alongside me. And, maybe most important, you were a kind friend
during hard times. Thank you for your understanding, your camara-
derie, and for sharing your gifts. From the bottom of my bad-at-run-
on-sentences-and—love for an em dash—grateful heart.

Erin Kane,
Thank you for taking a chance on me. I picked you out of all my meetings and hoped you picked me too. When you did, I cried. My gut instinct was right; we make great partners. Your continued calming
presence, warmth, and immensely helpful edits, suggestions, guidance, and encouragement have made this book come to life. Thank
you for your patience while this green-behind-the-ears-first-time
author tried to understand the publishing industry. No question was
too silly! I'm so grateful that you were open to my suggestions, that

you pushed me just a little bit further than I thought I could go, and that you had understanding for the calamity of writing with kids and offered wiggle room for deadlines because of them. I've absolutely loved every moment of working with you.

My Teams,

Thank you to the wondrous teams at Ballantine Books, Lucinda Literary, and Penguin Random House for being champions and guides in getting *RHLW* out into the world. I am so grateful to all the people who have touched this book. Special thanks to Julia Colluci and Connor Eck, Sanyu Dillon, Kara Welsh, Kim Hovey, Jennifer Hershey, Robbin Schiff, Rachel Keuch, Paolo Pepe, Megan Whalen, Kathleen Quinlan, Diane Hobbing, Cindy Berman, and Gina Wachtel. Dan Blank for your guidance, Rich Wade for my amazing author photo, Carolyn Hawkins for unlocking my cover concept, and Lane Kendall for your talent and willingness. Lindsay Newton for scrubbing up my bio and helping with feedback so early on. Finally, thank you to Beth Fisher at Levine Greenberg Rostan Literary Agency, who I met on a hike and very kindly introduced me to my very first literary agent! I'm forever grateful to Lindsay Edgecombe and the LGR team for getting me started on my publishing journey.

To those who were integral in
helping me learn, think, or grow.

To the many, many authors, writers, researchers, and artists whose work was a source for these pages, I am so grateful for your craft, experiences, and expertise. I learned so much and I am forever in awe of the heart and soul put into the works. And to my countless inspirations publishing books of their own, podcasts, newsletters, or artists making stories come to life, thank you for filling my creative well.

For the people throughout the book who graced these pages in spirit or by alias, thank you for the incredible impact you've made in my life. If we've crossed paths in some way on this long journey to here, I'm grateful to you for being a part of my story.

To my *Folk Rebellion* community without whom I would not be here writing this book. Thank you for believing in the mission, the brand, the lifestyle, the media, and me. *Folk Rebellion* had some of the best readers, followers, fans, ambassadors, partners, community, and team members. I just want you to know I saw you. I saw all your emails, DM's, purchases, shares, love notes, and shows of support. Though I couldn't handle it all (in more ways than one), I am forever proud of what we made together. There were so many people who helped it come to life. I'm forever grateful to Ryan, Lindsay, Andreaa, Lexi, Sarah, Pippa, Jessica, Stefan, Sophie, Dan, Jeff, Schuyler, Rigel, Matt, Maya, Elena, Nico, Kafi, Jenna, Morgan, James, Brooke, Mary Beth, Jacki, Kristin-Marie, Meghan, Carolina, Jerald, Tom, Katie, Andrew, Danielle, David, Lisette, Heather, Jaclyn, Leigh, Goldie, Emilie, Natasha, Jon, Jacklyn, Jasmine, Marla, Brittany, Lyon, and countless others I cannot include here because I am too long-winded and we are not-surprisingly already over on page count. (Which does not shock any of the aforementioned.) Thank you for helping me see that I was better as a creator than a business owner and for offering me grace while I waded through a very dark and difficult period of my personal and professional life. Thank you to all the organizations and brands who supported or got involved in some way. Thank you to the people who contributed to the Dispatch, those who subscribed, and those who helped get it out there. Thank you to everyone who featured FR, invited me to speak or collaborate, or helped spread the word. Thank you to all the off-the-grid retreaters who trusted me to take them to the ends of the earth, and then offline. To all the "rebels" who supported the crazy idea that we should not let technology run our life, thank you for trusting in such a counterculture idea and making it mainstream.

To my marketing life people for teaching me all you know or letting me teach you. Thank you for giving me the chance to be creative. Thank you to anyone who put their faith in me to mold their businesses, brands, campaigns, ideas. I took tremendous pride in the work.

To my mentors for showing me the ropes. Mark, for sharing all your wisdom and teachings with me. Steve, for letting me osmose your brilliant mind. John, for opening me up to the world and how to see it. Suzy, for everything.

To the publications that published my writing. *Whalebone Magazine, Simplify Magazine, Commune, Wanderlust, Bustle, Elite Daily, Huffington Post, Thrive,* and *The Guardian* for giving my words a reader.

To my creative collaborators for making ideas come to life. Especially thank you to Brooke Pathakis, Daniel Stangl, and Ilie Mitaru for your work on "What Day Is It?"

To the troublemakers who work tirelessly to try to make the world a better place for the next generation, I am grateful for your strength and voice. Special thanks to All Tech is Human, Center for Human Technology, Digital Wellness Institute, Fairplay Childhood Beyond Brands, LetGrow, and Moms Demand Action.

To those who were helpful in fostering ideas,
creativity, and community.

To the Aegean Center for the Fine Arts, Brooklyn Writers Collective, Write Away Europe, Binders, Camp Grounded, AllSwell Creative, Rock Your Bliss, KGB Bar Reading Series, Franklin Park Readers Series, and the Brooklyn Writers Space for helping me find my voice, my craft, my colleagues.

Suzy Batiz,
Thank you for being my modern-day philosopher friend to whiff ideas, musings, and existential crises past. I could talk to you forever about happiness, fulfillment, love, creativity, energy, success, family, and on. I am so grateful for your friendship, your love, your light.

Ivan Cash,
The greatest compliment I ever received was when a cool-ass artist told me my newspaper reminded them of Adbusters. I've been trying to live up to the hype ever since. Thank you for being a source of constant inspiration.

Jonathan Fields,

Your book fell off the shelf and landed on my toe many years ago while I perused the bookstore. I still have the page dog-eared it opened to. You were my inspiration all the way back then when I thought about becoming an author. Thank you for opening your world to me.

Oliver Jeffers,

You gifted me the greatest thing a writer with children could ever ask for. Space. Thank you for offering up your studio as a place of refuge and inspiration. When I wanted to crawl into a hole and quit, I instead crawled a few blocks away and up a few flights of stairs. I owe you (and Sam) a pint!

Erin Loechner,

Someday we will tell the story to our children of how the online world (where we no longer live) brought us together and that's how they ended up living on an offline commune (for which they hate us). A kindred spirit in philosophies, parenting, and writing. Thank you for it all.

Cal Newport,

Though I was bummed to be cut in the edits of your book, in the grand scheme of things, fair trade for the connection made. Thank you for your sage advice, the intros, and thumbs-up on my emailed idea! Look, it's a book!

Molly Rosen,

You saved me by offering for me to come check out your writers' workshop on grief. Thank you for reading between the lines of the "single mama" group post and not only giving me a reason to leave the house but a place to let the rage out of my writing. Yes, I believe I am a writer now.

Elissa Joy Watts,

I love that we've never actually met in person and yet here you are in these pages. I think that says everything there is to say about your heart, kindness, and way with words. Thank you for being my creative buddy.

Courtney Zoffness,
It was like I had all the ingredients for the recipe but didn't have the bowl. Thank you for sitting and listening while I told you exactly what I was trying to do. "It's an apology letter." I'm so grateful to you for that coffee conversation.

To those I turn to as my sources of constant inspiration,
Mary Beth LaRue, Jacki Carr, Sara Petersen, Kate Baer, Brooke McAlary, Ruby Warrington, Heather Havrilesky, Jia Tolentino, Nora McIrney, Jessica DeFino, Anne Helen Peterson, Sherry Turkle, Jean Twenge, David Byrne, Amanda Montei, Jaron Lanier, David Sax, Jessica Grose, and all the writers who are in the Background Reading section.

To my third rooms over the years,
Bua, Bar Tabac, Daylight Donuts, Props, Poppys, Fare and Folk, Duo, Life Cafe, Cafe Pick Me Up, KGB Bar, The Spaniard, Abelines, The Black Flamingo, Doc Holidays, St. Dymphnas, Lucy's, and The Bird House for providing me safe havens from which I could imbibe, read, or write with good coffee, cocktails, and people I could eavesdrop on. And big love to all my bodega boys for one of my most favorite third room relationships. You've doled out countless compliments and smiles to me over the years, and now my kids. Special shout-out to the boys at Tropical Juice Bar!

To the places and people
who made my life.

Utica,
For my roots.

New York City,
For my wings.

Nicole and Shannon,
I wouldn't want to know what life would be like without you two in it. It would be very boring for starters. You are my chosen family. I love

you more than words can say. No matter what changes (zip codes, decades, relationships, careers, fashion), our friendship doesn't. Thank you for it ALL. Nicole, I would bury the body for you. Shannon, I will go wherever you want. Yes, even when we are old and gray. Charlie's Angels for life. A blonde, a brunette, and a redhead walk into a bar . . .

Tracy,
I'm so grateful for our (now) three decades-long friendship. I love you for always being up for some trouble, countless conversations that inspired so many of these stories, and loving my kids even though, well, they're kids. You're one of the smartest people I know. I love your sappy ways.

Kristen,
I will forever be grateful to New York City for our meetcute. Can't remember the guys or the bar, but I remember that when we laughed together it was magic. Thank you for the adventures and always being up for some deep, honest conversation, or some high jinks.

315,
Tara, Lisa, Adrienne, Lindsey, and Mary, for the countless stories that can never actually go in a book. Jerry, I love you, kid. Jeff, for the dancing. Deb C., for celebrating me every step of the way. Mostly, thank you for the booties. Tom and the rest of the G-Men, for bringing me along. Davis, for my first adventure. To those who've come and gone through the different versions and phases of our lives but left behind nothing but love and lasting memories, thank you.

Red Rock and Bro J's fam,
I find it hard to be friends with anyone who's never worked in a bar or restaurant. To this day, you all remain some of the best, most interesting, inspiring, talented, hard-working, successful, silver-tongued, and quick-witted people I've ever known. I judge everyone else accordingly. You know who you are. Misfits for life—especially Joey & Drew.

Máire, Sarah, Carey, and the rest of the fam,
I am forever grateful for the love and support, kind words, copyediting, and grace given while I went through this very long process! Words don't suffice for everything you've done for our "wee" Beau and us. Thank you for welcoming me (and Hays) into your family with open arms. Máire, thank you especially for your visits, even though you must cross a pond. We love you.

Hays's village,
My fellow moms, it's an honor. Thank you to Allie, Anna, Carrie, Charlotte, Cinzia, Claire, Courtney, Ellen, Faye, Lauren, Liz, Liz, Lisa, Lise, Melissa, Toni, Sarah, Sophie, Stephanie, Virginie, and anyone else who dropped off the group text but remains in my heart. Happy to be past the pandemic-parenting years and excited for the middle school scaries alongside you. The Lima's, for our beautiful home. BunBun and SissyPants, for being my extended modern family in-laws and always being there no matter what. G, for giving me the greatest gift of becoming a mother. Forever grateful to all the grandparents, teachers, coaches, camp counselors, doctors, babysitters, bonus people responsible for helping raise a good human in the world.

Beau's village,
I am forever indebted to the incredible team at Hassenfeld's Children's Hospital who are responsible for Beau's heart and his ongoing care. Thank you to everyone who has touched his heart and ours, from the nurses, doctors, and surgeons to the techs, front desk staff, and security. You became our home away from home and extended family. Thank you for the care, love, and support you provided him but also the rest of us. We will see you all for #3 after the book tour! To all the people from Hays's village who have become a part of Beau's: Mickey, Stacey, Nick, Claire, Kyle, Jen, Shannon, Lauren, Simon, Mark, Maria, PJ, Archana, Kimberly, Sean, Stu, Rog, Chris, and the lot, plus the rest of our family and friends who have helped Beau by helping us, we are eternally grateful. A very special debt of gratitude is owed to Grandma and Papi for being our "other two passes." The coffees, food, big smiles, and positive energy you brought to the hospital room and our home got us through. And thank you to

Lisa and Liz for finding baby formula in our dystopian nightmare come true.

Siblings,

I've written this whole book with your voices in my head. Of course, most of it was ribbing. We are a rare breed, a small clan, a group, a gang, a pack. I can't imagine what life would've been like if I'd have gotten my wish every time I said I didn't want another brother or sister. Wishes thankfully ignored, I got three legends. Thank you for being my first friends, and later in life, my chosen friends. Thank you for all our inside jokes, secret languages, memories, adventures, nicknames, and even fights. It's because of you three that I have any character, fashion sense, or money at all. I'm sure you'll find many things to make fun of me for, but I'll know it's just your way of showing love. Andy, my handsome and hysterical little bro who shines brighter than anyone in a room. Hilarie, my witty and beautiful middle sis, the one to crack with the reward being the biggest heart of us all. Halley, my enterprising baby sis who never changes, whose kindness and smarts outweigh her stylish looks. Growing up with you has been one of the greatest happinesses of my life. Even though you're a bunch of assholes. Forever your Oliver.

Dad,

When I went out into the world, you showed me how to be wary of it, not afraid of it, and I am so grateful for that. You've somehow managed to teach me street smarts and skepticism, the perfect cocktail for the making of this book. You can keep sending me articles about what I should know; a lot of my research came from you! Dr. Doom paid off! Thank you for being such a big part of my life and the kids' lives. I'd have been lost without all the support you've given me over the past few years. This book wouldn't have had a chance if it wasn't for you jumping in to help with Beau and Hays. Thank you for the ideas, the time, the celebrations. I love you, Dad. We've come a long way. I'll call you later, okay? :)

Mom,

Hello Mother! (I know you read that in our voice.) If I typed up every card, letter, and inscribed book you've given me throughout my life (yes, of course I saved them) and put them all together, it would become a bestselling book on how to be a mother, how to love, and how to write. I am only me because of you. Thank you for my happy childhood, for making it *the* house and making it a home, for mending my broken hearts, for letting me be free to fail, and for always putting us kids first. You did it, Mom; all your hard work paid off. We are okay. Thank you for teaching me what is important in life. Author daughter also wants to say thank you for making me a reader at a young age, looking up cover designs, clipping newspaper articles about how to market books on social media, reading my words, believing my memories, and letting me know you were proud. Now, doggo, did you send a letter to Stephen King to tweet about your daughter's book? I love you more than a pig's eye. Do you hair me?

My children,
Beau, you are second to none, and I love you to the moon and back. I can't wait to see how you enjoy your precious life. Hays, my first love. I can't wait to see how far up, or far out, you go. I love you more than the moon and the stars. You both have forever changed me by becoming your mom/mama. Thank you for all the magic you bring into my life. I hope someday you can see that this was all for you. Go out into the world and use those big brains and bigger hearts to make big trouble. I love love love love you, Beau and Hays, my fire and ice. Yes, even more than pizza.

Kylian,
You literally never left my side. For being the best protector of my feet, the stoop, and Beau and Hays. Your loyalty and fluffy warm fur are greatly appreciated. RIP Tyson, who was just as loyal and a little more stinky.

In honor of my grandmothers, Margaret Thomas Clonan Regetz and Rose Arcuri Elefante Rescigno, who showed me the importance of strength, courage, and cheating at cards.

Background Reading

Raising Hell, Living Well is a culmination of thirty-six years of reading, forty-four years of living, and twenty-nine years of thinking and writing. These are the books and writers that have been enormously influential in the making of this book, and me.

Part I Under the Influence

Ariely, Dan. *Predictably Irrational: The Hidden Forces That Shape Our Decisions.* Rev. and expanded ed. HarperCollins, 2010.

Cialdini, Robert B., PhD. *Influence: The Psychology of Persuasion.* HarperCollins, 2009.

Graeber, David, and David Wengrow. *The Dawn of Everything: A New History of Humanity.* Penguin UK, 2021.

Heinrichs, Jay. *Thank You for Arguing: What Aristotle, Lincoln, and Homer Simpson Can Teach Us About the Art of Persuasion.* Three Rivers Press, 2013.

Parker, Priya. *The Art of Gathering: How We Meet and Why It Matters.* Penguin, 2020.

Plato. *The Republic.* Penguin UK, 2007.

Tracks

Bean, Philip A. *The Urban Colonists: Italian American Identity and Politics in Utica, New York.* Syracuse UP, 2010.

———. *La Colonia: Italian Life and Politics in Utica, New York, 1860–1960.* Utica College, Ethnic Heritage Studies Center, 2004.

Eberhardt, Jennifer L., PhD. *Biased: Uncovering the Hidden Prejudice That Shapes What We See, Think, and Do.* Penguin, 2020.

Harari, Yuval Noah. *Sapiens: A Brief History of Humankind.* Random House, 2014.

Hartman, Susan. *City of Refugees: The Story of Three Newcomers Who Breathed Life into a Dying American Town.* Beacon Press, 2022.

Johnson, Allan G. *Privilege, Power, and Difference.* McGraw-Hill Humanities Social, 2006.

Klosterman, Chuck. *The Nineties: A Book.* Penguin, 2022.

Leman, Kevin. *The Birth Order Book: Why You Are the Way You Are.* Revell, 2015.

Lewis, Michael. *The Undoing Project: A Friendship That Changed Our Minds.* W. W. Norton, 2016.

Morrison, Toni. *The Bluest Eye.* Washington Square Press, 1972.

Moskowitz, P. E. *How to Kill a City: Gentrification, Inequality, and the Fight for the Neighborhood.* Nation Books, 2018.

Pease, Bob. *Undoing Privilege: Unearned Advantage in a Divided World.* Bloomsbury, 2010.

Potocki, Rodger. *From the Inside: 40 Years of Reflections on Government, Politics, and Events in the Utica-Rome Area.* Infinity, 2010.

Zielbauer, Paul. "Looking to Prosper as a Melting Pot; Utica, Long in Decline, Welcomes an Influx of Refugees." *The New York Times,* May 7, 1999.

Picket Fences

Bolles, Richard N. *What Color Is Your Parachute? 2016: A Practical Manual for Job-Hunters and Career-Changers.* Ten Speed Press, 2015.

Bruni, Frank. *Where You Go Is Not Who You'll Be: An Antidote to the College Admissions Mania.* Grand Central, 2015.

Deresiewicz, William. *Excellent Sheep: The Miseducation of the American Elite and the Way to a Meaningful Life.* Simon and Schuster, 2015.

Kerouac, Jack. *On the Road.* Penguin, 1976.

Marriner, Mike, and Nathan Gebhard. *Roadtrip Nation: A Guide to Discovering Your Path in Life.* Ballantine Books, 2006.

Newland, Laura. *Chasing Zeroes: The Rise of Student Debt, the Fall of the College Ideal, and One Overachiever's Misguided Pursuit of Success.* Stonehall Press, 2013.

O'Meara, Rachael. *Pause: Harnessing the Life-Changing Power of Giving Yourself a Break.* National Geographic Books, 2017.

Pirsig, Robert M. *Zen and the Art of Motorcycle Maintenance: An Inquiry into Values.* HarperCollins, 2009.

Redfield, James, and Carol Adrienne. *The Celestine Prophecy: An Experiential Guide.* Random House, 1995.

Roadtrip Nation. *Roadmap: The Get-It-Together Guide for Figuring Out What to Do with Your Life.* Chronicle Books, 2015.

Twain, Mark. *The Adventures of Tom Sawyer.* Grosset and Dunlap, 1920.

Customs

Alighieri, Dante. *The Divine Comedy.* Penguin, 2003.

Brooks, Arthur C. "How to Want Less." *The Atlantic,* February 9, 2022.

Buettner, Dan. *The Blue Zones: 9 Lessons for Living Longer from the People Who've Lived the Longest.* 2nd ed. National Geographic Books, 2012.

Byrne, David. *Bicycle Diaries.* Penguin, 2010.

Coelho, Paulo. *The Alchemist.* Harper Thorsons, 2015.

Dweck, Carol S. *Self-Theories: Their Role in Motivation, Personality, and Development.* Psychology Press, 1999.

Gilbert, Elizabeth. *Eat, Pray, Love: One Woman's Search for Everything.* A&C Black, 2007.

Hampl, Patricia. *The Art of the Wasted Day.* Penguin, 2018.

Hodgkinson, Tom. *How to Be Idle: A Loafer's Manifesto.* HarperCollins, 2013.

Kamp, David. "The Trials and Triumphs of Santorini's Atlantis Books . . . New Threat to Its Existence." *Vanity Fair,* November 8, 2016.

Kazantzakis, Nikos. *Zorba the Greek.* Simon and Schuster, 1996.

Onstad, Katrina. *The Weekend Effect: The Life-Changing Benefits of Taking Time Off and Challenging the Cult of Overwork.* HarperOne, 2017.

Russell, Bertrand. *In Praise of Idleness and Other Essays.* Psychology Press, 2004.

Schlosser, Eric. *Fast Food Nation: The Dark Side of the All-American Meal.* Houghton Mifflin Harcourt, 2012.

Solnit, Rebecca. *Wanderlust: A History of Walking.* Viking Adult, 2000.

Stewart, Chris. *Driving over Lemons: An Optimist in Spain.* Vintage, 2001.

Tolle, Eckhart. *A New Earth: Awakening to Your Life's Purpose.* Penguin, 2006.

Weiner, Eric. *The Geography of Bliss: One Grump's Search for the Happiest Places in the World.* Twelve, 2009.

Bitter End

Astor, Maggie. "How the Politically Unthinkable Can Become Mainstream." *The New York Times,* February 26, 2019.

Battenfield, Jackie. *The Artist's Guide: How to Make a Living Doing What You Love.* Da Capo Lifelong Books, 2009.

Botton, Sari. *Never Can Say Goodbye: Writers on Their Unshakable Love for New York.* Simon and Schuster, 2014.

Bushnell, Candace. *Sex and the City.* HBO, 1998.

Calhoun, Ada. *St. Marks Is Dead: The Many Lives of America's Hippest Street.* National Geographic Books, 2016.

Crosley, Sloane. *I Was Told There'd Be Cake.* Portobello Books, 2012.

Graeber, David. *Bullshit Jobs: A Theory.* Simon and Schuster, 2019.

Guillebeau, Chris. *The Art of Non-conformity: Set Your Own Rules, Live the Life You Want, and Change the World.* Penguin, 2010.

Laing, Olivia. *The Lonely City: Adventures in the Art of Being Alone.* Macmillan, 2016.

Newman, Kristin. *What I Was Doing While You Were Breeding: A Memoir.* Crown, 2014.

Plath, Sylvia. *The Bell Jar.* Everyman's Library, 1963.

White, E. B. *Here Is New York.* New York Review of Books, 2011.

Whitehead, Colson. *The Colossus of New York.* Hachette UK, 2018.

Winter, Barbara. *Making a Living Without a Job: Winning Ways for Creating Work That You Love.* Bantam, 2009.

Part II How "They" Win Friends & Influence People

Barber, Benjamin R. *Consumed: How Markets Corrupt Children, Infantilize Adults, and Swallow Citizens Whole.* National Geographic Books, 2008.

Blas, Javier, and Jack Farchy. *The World for Sale: Money, Power, and the Traders Who Barter the Earth's Resources.* Random House, 2021.

Busse, Ryan. "The Gun Industry Created a New Consumer. Now It's Killing Us." *The Atlantic,* July 29, 2022.

Carnegie, Dale, and Dorothy Carnegie. *How to Win Friends and Influence People.* Turtleback, 1998.

Giblin, Rebecca, and Cory Doctorow. *Chokepoint Capitalism: How Big Tech and Big Content Captured Creative Labor Markets and How We'll Win Them Back.* Beacon Press, 2022.

Godin, Seth. *All Marketers Are Liars: The Underground Classic That Explains How Marketing Really Works—and Why Authenticity Is the Best Marketing of All.* National Geographic Books, 2012.

Kelley, Lora. "The Dark Side of Box Tops for Education." *The Atlantic,* April 16, 2021.

Klein, Naomi. *This Changes Everything: Capitalism vs. the Climate.* Simon and Schuster, 2015.

Linn, Susan. *Who's Raising the Kids? Big Tech, Big Business, and the Lives of Children.* New Press, 2022.

Palahniuk, Chuck. *Fight Club: A Novel.* W. W. Norton, 2005.

Sandel, Michael J. *What Money Can't Buy: The Moral Limits of Markets.* Farrar, Straus and Giroux, 2012.

Schor, Juliet B. *The Overspent American: Upscaling, Downshifting, and the New Consumer.* Basic Books, 1998.

Smith, Adam. *The Theory of Moral Sentiments.* Penguin, 2010.

Thaler, Richard H. *Misbehaving: The Making of Behavioral Economics.* W. W. Norton, 2015.

Watts, Alan W. *Does It Matter? Essays on Man's Relation to Materiality.* New World Library, 2010.

What's Your Favorite Color?

Chestnut, Beatrice, and Uranio Paes. *The Enneagram Guide to Waking Up: Find Your Path, Face Your Shadow, Discover Your True Self*. Red Wheel/ Weiser, 2021.

Fields, Jonathan. *Sparked: Discover Your Unique Imprint for Work That Makes You Come Alive*. HarperCollins, 2021.

Goffman, Erving. *The Presentation of Self in Everyday Life*. Viking Press, 1973.

Hartman, Taylor. *The People Code: It's All About Your Innate Motive*. Scribner, 2007.

Herman, Todd. *The Alter Ego Effect: The Power of Secret Identities to Transform Your Life*. HarperCollins, 2019.

Keirsey, David. *Please Understand Me II: Temperament, Character, Intelligence*. Prometheus Nemesis, 1998.

King, Patrick. *Read People Like a Book: How to Analyze, Understand, and Predict People's Emotions, Thoughts, Intentions, and Behaviors*. PKCS Media, 2020.

Kroeger, Otto, and Janet M. Thuesen. *Type Talk: The 16 Personality Types That Determine How We Live, Love, and Work*. Dell, 2013.

O'Neil, Cathy. *Weapons of Math Destruction: How Big Data Increases Inequality and Threatens Democracy*. Penguin, 2017.

Rath, Tom and Gallup. *StrengthsFinder 2.0*. Simon and Schuster, 2007.

Rohm, Robert A. *Positive Personality Profiles: D-I-S-C-Over Personality Insights to Understand Yourself . . . and Others!* Voyages Press, 1994.

Rubin, Gretchen. *The Four Tendencies: The Indispensable Personality Profiles That Reveal How to Make Your Life Better (and Other People's Lives Better, Too)*. Harmony, 2017.

Ziglar, Zig. *Selling 101: What Every Successful Sales Professional Needs to Know*. Thomas Nelson, 2003.

——. *Zig: The Autobiography of Zig Ziglar*. WaterBrook, 2002.

Zuboff, Shoshana. *The Age of Surveillance Capitalism: The Fight for a Human Future at the New Frontier of Power*. PublicAffairs, 2018.

Make Believe

Ariely, Dan. *Predictably Irrational: The Hidden Forces That Shape Our Decisions*. Rev. and expanded ed. HarperCollins, 2010.

Arrington, Tracy, and Matthew Frederick. *101 Things I Learned® in Advertising School*. Crown, 2018.

Burg, Bob. *The Art of Persuasion: Winning Without Intimidation*. Sound Wisdom, 2011.

Dobelli, Rolf. *The Art of Thinking Clearly*. HarperCollins, 2014.

Greene, Robert. *The 48 Laws of Power*. Penguin, 2000.

Heath, Chip, and Dan Heath. *Made to Stick: Why Some Ideas Survive and Others Die*. National Geographic Books, 2007.

Hendrix, Malcom. *Why Everyone Is Selling Fear: The Psychology of Targeting Pain-Points for Advertising*. Independently published, 2020.

Hutson, Matthew. *The 7 Laws of Magical Thinking: How Irrational Beliefs Keep Us Happy, Healthy, and Sane*. National Geographic Books, 2013.

Kay, Jane Holtz. *Asphalt Nation: How the Automobile Took Over America and How We Can Take It Back*. University of California Press, 1998.

Kuang, Cliff, and Robert Fabricant. *User Friendly: How the Hidden Rules of Design Are Changing the Way We Live, Work, and Play*. Random House, 2019.

Orwell, George. *1984*. HarperCollins, 2013.

Pentland, Alex. "Beyond the Echo Chamber." *Harvard Business Review*, November 2013.

Pomerantsev, Peter. *This Is Not Propaganda: Adventures in the War Against Reality*. Faber and Faber, 2019.

Rose, Todd. *Collective Illusions: Conformity, Complicity, and the Science of Why We Make Bad Decisions*. Hachette Go, 2022.

Skinner, B. F. *About Behaviorism*. Vintage, 1974.

Stanley, Jason. *How Propaganda Works*. Princeton UP, 2015.

Stephens-Davidowitz, Seth. *Everybody Lies*. Bloomsbury, 2017.

Thaler, Richard H., and Cass R. Sunstein. *Nudge: Improving Decisions About Health, Wealth, and Happiness*. Yale UP, 2009.

Heads Bowed, Eyes Closed

Aziz, Afdhel, and Bobby Jones. *Good Is the New Cool: Market Like You Give a Damn*. Simon and Schuster, 2016.

Berger, Jonah. *Contagious: Why Things Catch On*. Simon and Schuster, 2016.

Brown, Eliot, and Maureen Farrell. *The Cult of We: WeWork and the Great Start-Up Delusion*. HarperCollins UK, 2021.

Girard, Joe, and Stanley H. Brown. *How to Sell Anything to Anybody*. Simon and Schuster, 2006.

Godin, Seth. *Purple Cow: Transform Your Business by Being Remarkable*. Penguin Books Limited, 2005.

Goodson, Scott, and Chip Walker. *Activate Brand Purpose: How to Harness the Power of Movements to Transform Your Company*. Kogan Page, 2021.

Kearney Consumer Institute. "Have a Little Faith in Me: The Truth About Consumer Trust." May 2021.

Kornberger, Martin. *Brand Society: How Brands Transform Management and Lifestyle*. Cambridge UP, 2010.

Miller, Donald. *Building a StoryBrand: Clarify Your Message So Customers Will Listen*. HarperCollins Leadership, 2017.

Montell, Amanda. *Cultish: The Language of Fanaticism*. HarperCollins, 2021.

Ramaswamy, Vivek. *Woke, Inc.: Inside Corporate America's Social Justice Scam*. Hachette UK, 2021.

Read Kearney: "The Truth About Consumer Trust." *Kearney,* 2021.

Sacks, Oliver. *The Man Who Mistook His Wife for a Hat*. Picador, 2015.

Sheehan, Brian. *Loveworks: How the World's Top Marketers Make Emotional Connections to Win in the Marketplace*. powerHouse Books, 2013.

Shragai, Naomi. *The Man Who Mistook His Job for His Life: How to Thrive at Work by Leaving Your Emotional Baggage Behind*. Random House, 2021.

Sinek, Simon. *Start with Why: The Inspiring Million-Copy Bestseller That Will Help You Find Your Purpose*. Penguin UK, 2011.

Tracy, Brian. *The Psychology of Selling: How to Sell More, Easier, and Faster Than You Ever Thought Possible*. Thomas Nelson, 2006.

Truth in Advertising. "Companies Accused of Greenwashing." October 26, 2022.

Van Praet, Douglas. *Unconscious Branding: How Neuroscience Can Empower (and Inspire) Marketing*. Palgrave Macmillan, 2014.

Megaphones & Marshmallows

Berger, John. *Ways of Seeing*. Penguin UK, 2008.

Calarco, Jessica McCrory. "Why Rich Kids Are So Good at the Marshmallow Test." *The Atlantic,* April 28, 2021.

Cecconi, Christian, et al. "Schadenfreude: Malicious Joy in Social Media Interactions." *Frontiers in Psychology,* Frontiers Media, November 2020.

Coupland, Douglas. *Generation X: Tales for an Accelerated Culture*. Macmillan, 1991.

Eisenstein, Elizabeth L. *The Printing Press as an Agent of Change*. Cambridge UP, 1980.

Hornby, Nick. *High Fidelity: A Novel*. Riverhead Books, 2000.

Logan, Megan. "Psychology and TV: How Reality Programming Impacts Our Brains." *Inverse,* April 7, 2016.

McLuhan, Marshall. *Understanding Media: The Extensions of Man*. Routledge, 2001.

Mischel, Walter. *The Marshmallow Test: Understanding Self-Control and How to Master It*. Random House, 2014.

Nabi, Robin L., et al. "Reality-Based Television Programming and the Psychology of Its Appeal." *Media Psychology,* Taylor and Francis, November 2003.

Nussbaum, Emily. *I Like to Watch: Arguing My Way Through the TV Revolution*. Random House, 2019.

Postman, Neil. *Amusing Ourselves to Death: Public Discourse in the Age of Show Business*. Penguin, 2005.

Reiss, Steven P., and James Wiltz. "Why People Watch Reality TV." *Media Psychology,* Taylor and Francis, November 2004.

Resnick, Brian. "The 'Marshmallow Test' Said Patience Was a Key to Success. A New Replication Tells Us S'more." *Vox,* June 6, 2018.

Ritchie, Hannah. "Causes of Death." *Our World in Data,* February 14, 2018.

Saunders, George. *The Braindead Megaphone.* Penguin, 2007.

Sax, David. *The Revenge of Analog: Real Things and Why They Matter.* PublicAffairs, 2016.

Weirdos

Agrawal, Radha. *Belong: Find Your People, Create Community, and Live a More Connected Life.* Workman, 2018.

Bluestone, Gabrielle. *Hype: How Scammers, Grifters, Con Artists and Influencers Are Taking Over the Internet—and Why We're Following.* HarperCollins, 2021.

Bradberry, Travis, and Jean Greaves. *Emotional Intelligence 2.0.* TalentSmart, 2009.

Covey, Stephen R. *The 7 Habits of Highly Effective People: Powerful Lessons in Personal Change.* Simon and Schuster, 2013.

Craig, Andy, and Dave Yewman. *Weekend Language: Presenting with More Stories and Less PowerPoint.* Dash Consulting, 2013.

Gino, Francesca. *Rebel Talent: Why It Pays to Break the Rules at Work and in Life.* HarperCollins, 2018.

Gladwell, Malcolm. *Talking to Strangers: What We Should Know About the People We Don't Know.* Hachette UK, 2019.

Grant, Adam. *Originals: How Non-conformists Change the World.* Random House, 2016.

Junger, Sebastian. *Tribe: On Homecoming and Belonging.* Twelve, 2016.

Kishimi, Ichiro, and Fumitake Koga. *The Courage to Be Disliked: The Japanese Phenomenon That Shows You How to Change Your Life and Achieve Real Happiness.* Atria, 2018.

Klein, Gary. *The Power of Intuition: How to Use Your Gut Feelings to Make Better Decisions at Work.* Currency, 2007.

Li, Leon, et al. "Young Children Conform More to Norms Than to Preferences." *PLOS ONE,* Public Library of Science, 2021.

Stanier, Michael Bungay. *The Coaching Habit: Say Less, Ask More and Change the Way You Lead Forever.* Box of Crayons Press, 2016.

Theis, Brooke. "Is Elizabeth Holmes' Deep Voice a Power Move?," *Harper's Bazaar,* March 30, 2022.

Twenge, Jean M. *iGen: Why Today's Super-Connected Kids Are Growing Up Less Rebellious, More Tolerant, Less Happy—and Completely Unprepared for Adulthood—and What That Means for the Rest of Us.* Simon and Schuster, 2017.

Wilson, Eric G. *How to Be Weird: An Off-Kilter Guide to Living a One-of-a-Kind Life.* Penguin, 2022.

Part III The Bad Influence

"Art of Persuasion Is Getting Complex. Americans Are Bombarded Daily by People Trying to Influence Decisions." *Los Angeles Times,* July 23, 1992.

Holiday, Ryan. *Ego Is the Enemy.* Penguin, 2016.

————. *Trust Me, I'm Lying: Confessions of a Media Manipulator.* Penguin, 2012.

Kendi, Ibram X. *How to Raise an Antiracist.* One World, 2022.

Pratkanis, Anthony R., and Elliot Aronson. *Age of Propaganda: The Everyday Use and Abuse of Persuasion.* Macmillan, 2001.

Prinstein, Mitch. *Popular: Why Being Liked Is the Secret to Greater Success and Happiness.* Random House, 2017.

Pony Up

Alter, Adam. *Irresistible: The Rise of Addictive Technology and the Business of Keeping Us Hooked.* Penguin, 2017.

Bourdain, Anthony. *Kitchen Confidential.* A&C Black, 2013.

Burns, Robert. *A Night Out with Robert Burns: The Greatest Poems.* Canongate Books, 2009.

Carr, Nicholas. *The Shallows: What the Internet Is Doing to Our Brains.* W. W. Norton, 2010.

Danler, Stephanie. *Sweetbitter.* Knopf, 2019.

Eyal, Nir. *Hooked: How to Build Habit-Forming Products.* Penguin, 2014.

Grisel, Judith. *Never Enough: The Neuroscience and Experience of Addiction.* Scribe, 2019.

Lembke, Anna. *Dopamine Nation: Finding Balance in the Age of Indulgence.* Penguin, 2021.

Lieberman, Daniel Z., and Michael E. Long. *The Molecule of More: How a Single Chemical in Your Brain Drives Love, Sex, and Creativity—and Will Determine the Fate of the Human Race.* BenBella Books, 2019.

Maté, Gabor. *In the Realm of Hungry Ghosts: Close Encounters with Addiction.* Random House, 2018.

Roberts, Kevin. *Cyber Junkie: Escape the Gaming and Internet Trap.* Simon and Schuster, 2010.

Sapolsky, Robert M. *Behave: The Biology of Humans at Our Best and Worst.* Penguin, 2018.

Schoen, Marc. *Your Survival Instinct Is Killing You: Retrain Your Brain to Conquer Fear and Build Resilience.* National Geographic Books, 2014.

Weinschenk, Susan. *Neuro Web Design: What Makes Them Click?* New Riders, 2018.

————. *How to Get People to Do Stuff: Master the Art and Science of Persuasion and Motivation.* New Riders, 2013.

Good on Paper

Konnikova, Maria. *The Confidence Game: Why We Fall for It . . . Every Time.* Penguin, 2017.

Lewis, Michael. *The Big Short: Inside the Doomsday Machine.* National Geographic Books, 2010.

Lorenz, Taylor. "On The Internet, No One Knows You're Not Rich. Except This Account." *The New York Times,* November 12, 2019.

Monroe, Rachel. "I'm a Life Coach, You're a Life Coach: The Rise of an Unregulated Industry." *The Guardian,* December 5, 2022.

Payne, Keith. *The Broken Ladder: How Inequality Changes the Way We Think, Live, and Die.* Hachette UK, 2017.

Semuels, Alana. "The Men Peddling the 'Secrets' to Getting Rich on Amazon." *The Atlantic,* March 8, 2021.

Wood, Matthew J. A., et al. "Fake It 'Til You Make It: Hazards of a Cultural Norm in Entrepreneurship." *Business Horizons,* Elsevier, December 2021.

Something Borrowed

Burkus, David. *Friend of a Friend . . . : Understanding the Hidden Networks That Can Transform Your Life and Your Career.* HarperCollins, 2018.

Christakis, Nicholas A., and James H. Fowler. *Connected: The Surprising Power of Our Social Networks and How They Shape Our Lives—How Your Friends' Friends' Friends Affect Everything You Feel, Think, and Do.* Back Bay Books, 2011.

Draper, Robert. "Is the Power Lunch Officially Dead?" *GQ,* September 2015.

Fussell, Paul. *Class: A Guide Through the American Status System.* Simon and Schuster, 1992.

Gonzalez, Xochitl. "The New Case for Social Climbing." *The Atlantic,* January 12, 2023.

Hadnagy, Christopher. *Social Engineering: The Art of Human Hacking.* John Wiley and Sons, 2010.

Kendall, Diana. *Members Only: Elite Clubs and the Process of Exclusion.* Rowman and Littlefield, 2008.

Laird, Pamela Walker. *Pull: Networking and Success Since Benjamin Franklin.* Harvard UP, 2006.

Program on Negotiation, Harvard Law School. "Borrowing Influence: How to Get By with a Little Help from Your Friends." 2010.

Rosy Future

Abad-Santos, Alex. "Gaslight Gatekeep Girlboss, Explained." *Vox,* 2021.

Cain, Susan. *Quiet: The Power of Introverts in a World That Can't Stop Talking.* Penguin UK, 2012.

Farber, Jim. "Bespoke This, Bespoke That. Enough Already." *The New York Times,* August 8, 2016.

Gladwell, Malcolm. *The Tipping Point: How Little Things Can Make a Big Difference.* Wheeler, 2003.

Goldberg, Emma. "What Sheryl Sandberg's 'Lean In' Has Meant to Women." *The New York Times,* June 2, 2022.

Higham, William. *The Next Big Thing: Spotting and Forecasting Consumer Trends for Profit.* Kogan Page, 2009.

Klosterman, Chuck. *Sex, Drugs, and Cocoa Puffs: A Low Culture Manifesto.* Faber and Faber, 2013.

Lanier, Jaron. *Who Owns the Future?* Simon and Schuster, 2014.

Mull, Amanda. "The Girlboss; and the Myth of Corporate Female Empowerment." *The Atlantic,* 2020.

Pang, Alex S. *Rest: Why You Get More Done When You Work Less.* Penguin UK, 2016.

Petersen, Anne Helen. *Can't Even: How Millennials Became the Burnout Generation.* Random House, 2021.

Salzman, Marian. *The New Megatrends: Seeing Clearly in the Age of Disruption.* Currency, 2022.

Sandberg, Sheryl. *Lean In: Women, Work, and the Will to Lead.* Random House, 2013.

Unplugged & Loved

Crook, Christina. *The Joy of Missing Out: Finding Balance in a Wired World.* New Society, 2014.

Deresiewicz, William. *The Death of the Artist: How Creators Are Struggling to Survive in the Age of Billionaires and Big Tech.* Henry Holt, 2020.

Eyal, Nir. *Indistractable: How to Control Your Attention and Choose Your Life.* Bloomsbury, 2019.

Flanders, Cait. *Adventures in Opting Out: A Field Guide to Leading an Intentional Life.* Hachette UK, 2020.

Gilbert, Elizabeth. *Big Magic: How to Live a Creative Life, and Let Go of Your Fear.* Bloomsbury, 2015.

Grace, Marlee. *How to Not Always Be Working: A Toolkit for Creativity and Radical Self-Care.* HarperCollins, 2018.

Havrilesky, Heather. *What If This Were Enough? Essays.* Anchor, 2018.

Holiday, Ryan. *Stillness Is the Key: An Ancient Strategy for Modern Life.* Profile Books, 2019.

Jarvis, Paul. *Company of One: Why Staying Small Is the Next Big Thing for Business.* Houghton Mifflin Harcourt, 2019.

Kagge, Erling. *Silence: In the Age of Noise.* Vintage, 2018.

Loechner, Erin. *Chasing Slow: Courage to Journey Off the Beaten Path.* Zondervan, 2017.

McAlary, Brooke. *Slow: Simple Living for a Frantic World.* Sourcebooks, 2018.

McKeown, Greg. *Essentialism: The Disciplined Pursuit of Less*. Random House, 2014.

Nagoski, Emily, and Amelia Nagoski. *Burnout: The Secret to Solving the Stress Cycle*. Random Bloomsbury Publishing, 2015.

Newport, Cal. *Digital Minimalism: Choosing a Focused Life in a Noisy World*. Penguin, 2019.

———. *Deep Work: Rules for Focused Success in a Distracted World*. Hachette UK, 2006.

Pink, Daniel H. *Drive: The Surprising Truth About What Motivates Us*. Canongate Books, 2010.

Sax, David. *The Future Is Analog: How to Create a More Human World*. Hachette UK, 2021.

Skidelsky, Robert, and Edward Skidelsky. *How Much Is Enough? Money and the Good Life*. Other Press, 2013.

Solnit, Rebecca. "Rebecca Solnit on Women's Work and the Myth of the Art Monster." *Literary Hub*, 2019.

Stafford, Rachel Macy. *Hands Free Mama: A Guide to Putting Down the Phone, Burning the To-Do List, and Letting Go of Perfection to Grasp What Really Matters*. Zondervan, 2014.

Zuckerberg, Randi. *Dot Complicated: Untangling Our Wired Lives*. HarperOne, 2015.

Hanging Up

Bauerlein, Mark. *The Digital Divide: Arguments for and Against Facebook, Google, Texting, and the Age of Social Networking*. National Geographic Books, 2011.

Brynjolfsson, Erik, and Andrew McAfee. *The Second Machine Age: Work, Progress, and Prosperity in a Time of Brilliant Technologies*. W. W. Norton, 2014.

Camus, Albert. *The Myth of Sisyphus*. Penguin UK, 2013.

Durant, Will. *The Story of Philosophy*. Courier Dover, 2022.

Ford, Rob. *Web Design: The Evolution of the Digital World 1990–Today*. Taschen, 2019.

Hollins, Peter. *Think Like a Genius: How to Go Outside the Box, Analyze Deeply, Creatively Solve Problems, and Innovate*. PKCS Media, 2021.

Huffington, Arianna. *Thrive: The Third Metric to Redefining Success and Creating a Happier Life*. Random House, 2015.

Isaacson, Walter. *Steve Jobs*. Simon and Schuster, 2021.

Jasanoff, Sheila. *The Ethics of Invention*. National Geographic Books, 2016.

Konnikova, Maria. "What's Lost as Handwriting Fades." *The New York Times*, June 2, 2014.

Lanier, Jaron. *Ten Arguments for Deleting Your Social Media Accounts Right Now*. Henry Holt, 2018.

———. *You Are Not a Gadget: A Manifesto*. Penguin UK, 2010.

Murphy, Cullen. "What the Internet Can Learn from the Printing Press." *The Atlantic,* 2019.

Odell, Jenny. *How to Do Nothing: Resisting the Attention Economy.* Melville House, 2019.

"Opinion: The New Radicalization of the Internet." *The New York Times,* November 24, 2018.

Price, Catherine. *How to Break Up with Your Phone: The 30-Day Plan to Take Back Your Life.* Hachette UK, 2018.

Rushkoff, Douglas. *Present Shock: When Everything Happens Now.* Penguin, 2013.

Shlain, Tiffany. *24/6: The Power of Unplugging One Day a Week.* Simon and Schuster, 2019.

Tolentino, Jia. *Trick Mirror: Reflections on Self-Delusion.* Random House, 2019.

Turkle, Sherry. *Reclaiming Conversation: The Power of Talk in a Digital Age.* National Geographic Books, 2016.

Watts, Alan. *The Wisdom of Insecurity.* Vintage, 2011.

Wolf, Maryanne. *Reader, Come Home: The Reading Brain in a Digital World.* Harper, 2018.

Modern Bullshit

Ansari, Aziz, and Eric Klinenberg. *Modern Romance.* Penguin, 2015.

Bradfield, Damien. *The Trust Manifesto: What You Need to Do to Create a Better Internet.* Penguin UK, 2019.

Brown, Brené. *Braving the Wilderness: The Quest for True Belonging and the Courage to Stand Alone.* Random House, 2017.

Daum, Meghan. *The Problem with Everything: My Journey Through the New Culture Wars.* Simon and Schuster, 2019.

Davis, Allison P. "The Rise and Fall of Babe.net and the Aziz Ansari Story." The Cut, *New York,* June 23, 2019.

Didion, Joan. *Slouching Towards Bethlehem: Essays.* Open Road Media, 2017.

Grenny, Joseph, Kerry Patterson, Ron McMillan, Al Switzler, and Emily Gregory. *Crucial Conversations: Tools for Talking When Stakes Are High.* 3rd ed. McGraw Hill Professional, 2021.

Morrison, Toni. *Burn This Book: Notes on Literature and Engagement.* Harper Paperbacks, 2012.

Moss, Jeremiah. *Feral City: On Finding Liberation in Lockdown New York.* W. W. Norton, 2022.

O'Reilly, A. J. *The Martyrs of the Coliseum or Historical Records of the Great Amphitheater of Ancient Rome.* TAN Books, 2016.

Phillips, Whitney. *This Is Why We Can't Have Nice Things: Mapping the Relationship Between Online Trolling and Mainstream Culture.* MIT Press, 2015.

Ronson, Jon. *So You've Been Publicly Shamed*. Pan Macmillan, 2015.

Rushdie, Salman. *The Satanic Verses*. Random House, 2010.

Schulman, Sarah. *Conflict Is Not Abuse: Overstating Harm, Community Responsibility, and the Duty of Repair*. Arsenal Pulp Press, 2016.

Strickler, Yancey. "The Dark Forest Theory of the Internet—OneZero." *Medium*, December 10, 2021.

Wu, Tim. *The Attention Merchants: The Epic Struggle to Get Inside Our Heads*. Atlantic Books, 2017.

80's Mom

Beck, Richard. *We Believe the Children: A Moral Panic in the 1980s*. PublicAffairs, 2015.

Brooks, Kim. *Small Animals: Parenthood in the Age of Fear*. Flatiron Books, 2018.

Brown, Tracey, and Michael Hanlon. *In the Interests of Safety: The Absurd Rules That Blight Our Lives and How We Can Change Them*. Hachette UK, 2014.

Chan, Jessamine. *The School for Good Mothers: A Novel*. Simon and Schuster, 2022.

Curtis, Drew. *It's Not News, It's Fark: How Mass Media Tries to Pass Off Crap as News*. Penguin, 2007.

Druckerman, Pamela. *Bringing up Bébé: One American Mother Discovers the Wisdom of French Parenting (Now with Bébé Day by Day: 100 Keys to French Parenting)*. Penguin, 2014.

Furedi, Frank. *How Fear Works: Culture of Fear in the Twenty-first Century*. Bloomsbury, 2018.

Gardner, Daniel. *The Science of Fear: How the Culture of Fear Manipulates Your Brain*. National Geographic Books, 2009.

Hendricks, Gay. *The Big Leap: Conquer Your Hidden Fear and Take Life to the Next Level*. HarperCollins, 2009.

Holland, Eva. *Nerve: A Personal Journey Through the Science of Fear*. Penguin, 2020.

Levitt, Steven D., and Stephen J. Dubner. *Freakonomics: A Rogue Economist Explores the Hidden Side of Everything*. HarperCollins, 2005.

Louv, Richard. *Last Child in the Woods: Saving Our Children from Nature-Deficit Disorder*. Algonquin Books, 2008.

Lukianoff, Greg, and Jonathan Haidt. *The Coddling of the American Mind: How Good Intentions and Bad Ideas Are Setting Up a Generation for Failure*. Penguin, 2018.

Nezhukumatathil, Aimee. *World of Wonders: In Praise of Fireflies, Whale Sharks, and Other Astonishments*. Milkweed Editions, 2020.

Redleaf, Diane L. *They Took the Kids Last Night: How the Child Protection System Puts Families at Risk*. ABC-CLIO, 2018.

Renfro, Paul M. *Stranger Danger: Family Values, Childhood, and the American Carceral State*. Oxford UP, 2020.

Rosin, Hanna. "The Overprotected Kid." *The Atlantic,* November 2018.

Ryan, Christopher. *Civilized to Death: The Price of Progress.* Avid Reader Press, 2020.

Sinclair, Upton. *The Brass Check: A Study of American Journalism.* Cornell University LIbrary, 1920.

Skenazy, Lenore. *Free-Range Kids: How to Raise Safe, Self-Reliant Children (Without Going Nuts with Worry).* John Wiley and Sons, 2010.

Smith, Betty. *A Tree Grows in Brooklyn.* Random House, 1992.

Sutherland, Stuart. *Irrationality: The Enemy Within.* Pinter and Martin, 2013.

Zomorodi, Manoush. *Bored and Brilliant: How Spacing Out Can Unlock Your Most Productive and Creative Self.* St. Martin's Press, 2017.

Vulnerability Clickbait

Brown, Brené. *Daring Greatly: How the Courage to Be Vulnerable Transforms the Way We Live, Love, Parent, and Lead.* National Geographic Books, 2015.

Buettner, Dan. *The Blue Zone: 9 Lessons for Living Longer from the People Who've Lived the Longest.* 2nd ed. National Geographic Books, 2012.

Busch, Akiko. *How to Disappear: Notes on Invisibility in a Time of Transparency.* Penguin, 2019.

Cameron, J. *The Artist's Way: A Spiritual Path to Higher Creativity.* Penguin, 2020.

Camus, Albert. *Create Dangerously: The Power and Responsibility of the Artist.* National Geographic Books, 2019.

Didion, Joan. *The Year of Magical Thinking: A Play by Joan Didion Based on Her Memoir.* Vintage, 2007.

Duhigg, Charles. *The Power of Habit: Why We Do What We Do and How to Change.* Random House, 2012.

Longhurst, Erin Niimi. *A Little Book of Japanese Contentments: Ikigai, Forest Bathing, Wabi-sabi, and More.* Chronicle Books, 2018.

May, Katherine. *Wintering: The Power of Rest and Retreat in Difficult Times.* Penguin, 2020.

McIrney, Nora. *Bad Vibes Only.* Atria, 2022.

Ostaseski, Frank. *The Five Invitations: Discovering What Death Can Teach Us About Living Fully.* Pan Macmillan, 2017.

Petersen, Sara. *Momfluenced: Inside the Maddening, Picture-Perfect World of Mommy Influencer Culture.* Beacon Press, 2023.

Quindlen, Anna. *A Short Guide to a Happy Life.* National Geographic Books, 2001.

Scott, Laurence. *The Four-Dimensional Human: Ways of Being in the Digital World.* Random House, 2015.

Turkle, Sherry. *Alone Together: Why We Expect More from Technology and Less from Each Other.* Hachette UK, 2017.

Wake Up

Aamodt, Sandra, and Sam Wang. *Welcome to Your Brain: Why You Lose Your Car Keys but Never Forget How to Drive and Other Puzzles of Everyday Behavior.* Bloomsbury, 2009.

Adichie, Chimamanda Ngozi. *We Should All Be Feminists.* Vintage, 2014.

Banaji, Mahzarin R., and Anthony G. Greenwald. *Blindspot: Hidden Biases of Good People.* Bantam, 2016.

Doyle, Glennon. *Untamed.* National Geographic Books, 2020.

Estés, Clarissa Pinkola. *Women Who Run with the Wolves: Myths and Stories of the Wild Woman Archetype.* Ballantine Books, 1996.

Gay, Roxane. *Bad Feminist: Essays.* Harper Perennial, 2014.

Heath, Dan. *Upstream: The Quest to Solve Problems Before They Happen.* Avid Reader Press, 2020.

Holland, Jack. *A Brief History of Misogyny: The World's Oldest Prejudice.* Hachette UK, 2012.

hooks, bell. *The Will to Change: Men, Masculinity, and Love.* Washington Square Press, 2004.

Perez, Caroline Criado. *Invisible Women: Data Bias in a World Designed for Men.* Abrams Press, 2021.

Petersen, Anne Helen. *Too Fat, Too Slutty, Too Loud: The Rise and Reign of the Unruly Woman.* Penguin, 2017.

Solnit, Rebecca. *Men Explain Things to Me.* Haymarket Books, 2019.

Pick a Fight

Brooks, David. "The Nuclear Family Was a Mistake." *The Atlantic,* March 2022.

Chapman, Gary. *The Five Love Languages: How to Express Heartfelt Commitment to Your Mate.* Northfield Press, 2010.

Coontz, Stephanie. *The Way We Never Were: American Families and the Nostalgia Trap.* Hachette UK, 2016.

Denborough, David. *Retelling the Stories of Our Lives: Everyday Narrative Therapy to Draw Inspiration and Transform Experience.* W. W. Norton, 2014.

Edelman, Hope. *The AfterGrief: Finding Your Way Along the Long Arc of Loss.* Ballantine Books, 2020.

Febos, Melissa. *Body Work: The Radical Power of Personal Narrative.* Catapult, 2022.

Gottman, John, PhD., and Nan Silver. *The Seven Principles for Making Marriage Work: A Practical Guide from the Country's Foremost Relationship Expert.* Harmony, 2015.

Guzmán, Mónica. *I Never Thought of It That Way: How to Have Fearlessly Curious Conversations in Dangerously Divided Times.* BenBella Books, 2022.

Levine, Amir, and Rachel Heller. *Attached: The New Science of Adult Attachment and How It Can Help You Find—and Keep—Love.* Penguin, 2012.

Orloff, Judith. *The Empath's Survival Guide: Life Strategies for Sensitive People.* Sounds True, 2017.

Paltrow, Gwyneth. "From the Archive: Gwyneth Paltrow on Her Conscious Uncoupling Journey." *British Vogue,* September 27, 2022.

Rao, Ranjani. *Rewriting My Happily Ever After: A Memoir of Divorce and Discovery.* Story Artisan Press, 2021.

Smith, Emily Esfahani. "The Secret to Love Is Just Kindness." *The Atlantic,* September 20, 2022.

Thunberg, Greta. *No One Is Too Small to Make a Difference.* Exp. ed. National Geographic Books, 2020.

van der Kolk, Bessel A. *The Body Keeps the Score: Brain, Mind, and Body in the Healing of Trauma.* Penguin, 2015.

Band Together

Basu, Tanya. "How Google Docs Became the Social Media of the Resistance." *MIT Technology Review,* June 2020.

Bennett, Michael, and Sarah Bennett. *F*ck Feelings: One Shrink's Practical Advice for Managing All Life's Impossible Problems.* Simon and Schuster, 2015.

Han, Hahrie, Elizabeth McKenna, and Michelle Oyakawa. *Prisms of the People: Power and Organizing in Twenty-first-Century America.* University of Chicago Press, 2021.

Lipez, Zachary. "Woodstock '99 Was a Violent Disaster That Predicted America's Future." Vice, July 24, 2019.

Maslow, Abraham H. *A Theory of Human Motivation.* General Press, 2019.

Mina, An Xiao. *Memes to Movements: How the World's Most Viral Media Is Changing Social Protest and Power.* Beacon Press, 2019.

Murthy, Vivek H. *Together: The Healing Power of Human Connection in a Sometimes Lonely World.* HarperCollins, 2020.

Petrusich, Amanda. "Woodstock '99 and the Rise of Toxic Masculinity." *The New Yorker,* July 30, 2021.

Rushkoff, Douglas. *Team Human.* National Geographic Books, 2019.

Solnit, Rebecca. *A Paradise Built in Hell: The Extraordinary Communities That Arise in Disaster.* National Geographic Books, 2010.

Watts, Shannon. *Fight Like a Mother: How a Grassroots Movement Took on the Gun Lobby and Why Women Will Change the World.* HarperOne, 2020.

Make Trouble

Grant, Adam. *Originals: How Non-conformists Move the World*. Penguin, 2016.

Jones, Brian Jay. *Becoming Dr. Seuss: Theodor Geisel and the Making of an American Imagination*. Penguin, 2019.

Nader, Ralph. *Unsafe at Any Speed*. Grossman Publishers, 1966.

Popova, Maria. *Figuring*. Vintage, 2019.

Rayner, Cynthia, and François Bonnici. *The Systems Work of Social Change: How to Harness Connection, Context, and Power to Cultivate Deep and Enduring Change*. Oxford UP, 2021.

Rogers, Fred. *The World According to Mister Rogers: Important Things to Remember*. Hachette UK, 2003.

Shalaby, Carla. *Troublemakers: Lessons from Children Disrupting School*. New Press, 2017.

Tharp, Twyla. *The Creative Habit: Learn It and Use It for Life*. Simon and Schuster, 2009.

Vonnegut, Kurt. *Breakfast of Champions*. Random House, 1992.

Waters, John. *Make Trouble*. Hachette UK, 2017.

The "Plant Your Flag" Practice

Brooks, David. *The Road to Character*. Random House, 2015.

Chopra, Deepak, and Menas C. Kafatos. *You Are the Universe: Discovering Your Cosmic Self and Why It Matters*. Harmony, 2017.

Gibran, Kahlil. *The Prophet*. Deluxe ed. Knopf, 1951.

Hawking, Stephen. *Brief Answers to the Big Questions: The Final Book from Stephen Hawking*. Hachette UK, 2018.

Manson, Mark. *The Subtle Art of Not Giving a F*ck: A Counterintuitive Approach to Living a Good Life*. HarperCollins, 2016.

Otting, Laura Gassner. *Limitless: How to Ignore Everybody, Carve Your Own Path, and Live Your Best Life*. IdeaPress, 2020.

Peck, M. Scott. *The Road Less Traveled: A New Psychology of Love, Traditional Values, and Spiritual Growth*. 25th anniv. ed. Simon and Schuster, 2002.

Pressfield, Steven. *Put Your Ass Where Your Heart Wants to Be*. Sarsaparilla Media, 2022.

———. *The War of Art: Break Through the Blocks and Win Your Inner Creative Battles*. Black Irish Entertainment, 2012.

Singer, Michael A. *Living Untethered: Beyond the Human Predicament*. New Harbinger, 2022.

———. *The Surrender Experiment: My Journey into Life's Perfection*. Yellow Kite, 2016.

The "Peel the Onion" Technique

Bergstrom, Carl T., and Jevin D. West. *Calling Bullshit: The Art of Skepticism in a Data-Driven World*. Random House Trade Paperbacks, 2021.

Carter, Lee Hartley. *Persuasion: Convincing Others When Facts Don't Seem to Matter*. National Geographic Books, 2020.

Grant, Adam. *Think Again: The Power of Knowing What You Don't Know*. Random House, 2021.

Greif, Mark. *Against Everything: Essays*. Vintage, 2016.

Hari, Johann. *Stolen Focus: Why You Can't Pay Attention*. Bloomsbury, 2022.

Levitt, Steven D., and Stephen J. Dubner. *Think Like a Freak: The Authors of* Freakonomics *Offer to Retrain Your Brain*. William Morrow, 2014.

Moses, Michele. "The Truth About Selfie Culture." *The New Yorker,* April 30, 2018.

Storr, Will. *Selfie: How We Became So Self-Obsessed and What It's Doing to Us*. Abrams, 2019.

Wittgenstein, Ludwig. *On Certainty*. Wiley-Blackwell, 1991.

Zahariades, Damon. *The Art of Saying No: How to Stand Your Ground, Reclaim Time and Energy, and Refuse to Be Taken for Granted*. Independently published, 2017.

The "Middle Fish" Assessment

Heath, Chip, and Dan Heath. *The Power of Moments: Why Certain Experiences Have Extraordinary Impact*. Simon and Schuster, 2017.

hooks, bell. *All About Love: New Visions*. HarperCollins, 2018.

Ruiz, Don Miguel, and Janet Mills. *The Four Agreements: A Practical Guide to Personal Freedom*. Hay House, 1997.

Strickler, Yancey. *This Could Be Our Future: A Manifesto for a More Generous World*. Penguin, 2020.

Tolle, Eckhart. *The Power of Now: A Guide to Spiritual Enlightenment*. New World Library, 2010.

The "Snowman" Effect

Blum, Andrew. *The Weather Machine: How We See into the Future*. Random House, 2019.

Brown, Stuart L., and Christopher C. Vaughan. *Play: How It Shapes the Brain, Opens the Imagination, and Invigorates the Soul*. Avery Publishing Group, 2010.

Bukowski, Charles. *Mockingbird Wish Me Luck*. HarperCollins, 2009.

Clear, James. *Atomic Habits: An Easy and Proven Way to Build Good Habits and Break Bad Ones*. National Geographic Books, 2018.

Kabat-Zinn, Jon. *Wherever You Go, There You Are: Mindfulness Meditation in Everyday Life.* Hyperion, 2005.

Kempton, Beth. *Wabi Sabi: Japanese Wisdom for a Perfectly Imperfect Life.* Hachette UK, 2018.

Kleon, Austin. *Steal Like an Artist: 10 Things Nobody Told You About Being Creative.* Hachette UK, 2012.

Louv, Richard. *Vitamin N: The Essential Guide to a Nature-Rich Life.* Algonquin Books, 2017.

Maxwell, John C. *Intentional Living: Choosing a Life That Matters.* Center Street, 2015.

Oxenreider, Tsh. *Notes from a Blue Bike: The Art of Living Intentionally in a Chaotic World.* Thomas Nelson, 2014.

Williams, Florence. *The Nature Fix: Why Nature Makes Us Happier, Healthier, and More Creative.* W. W. Norton, 2017.

Notes can be found on jessicaelefante.com.

About the Author

Jessica Elefante is a writer and ~~bullshit~~ artist who has spent the last few decades examining what it means to be human in our modern world. Her essays have appeared in *The Guardian, The Huffington Post,* and more. As the founder of acclaimed *Folk Rebellion* and a critic of today's culture, Elefante's award-winning talks, films, and work have been featured by *Vogue,* the *Los Angeles Times, The Observer, Paper* magazine, *Wired,* and elsewhere. In her previous life as a brand strategist, she was recognized as one of Brand Innovators' 40 Under 40 and has been a guest lecturer at Columbia Business School and New York University. She's influenced by the social, cultural, and technological circumstances of her life but mostly by her desire to lead a colorful one. Raised in Upstate New York, she now lives in Brooklyn with her family. She is no longer bullshitting.

jessicaelefante.com
Instagram: @folkrebellion
Substack: Modern Bullshit

About the Type

This book was set in Sabon, a typeface designed by the well-known German typographer Jan Tschichold (1902–74). Sabon's design is based upon the original letterforms of sixteenth-century French type designer Claude Garamond and was created specifically to be used for three sources: foundry type for hand composition, Linotype, and Monotype. Tschichold named his typeface for the famous Frankfurt typefounder Jacques Sabon (c. 1520–80).